International Conference on the Frontiers in Atomic, Molecular, and Optical Physics (AMO 2016)

Journal of Physics: Conference Series Volume 752

Shanghai, China
23 – 26 May 2016

ISBN: 978-1-5108-3061-5
ISSN: 1742-6588

Printed from e-media with permission by:

Curran Associates, Inc.
57 Morehouse Lane
Red Hook, NY 12571

proceedings.com

Some format issues inherent in the e-media version may also appear in this print version.

Printed by Curran Associates, Inc. (2016)

For permission requests, please contact the Institute of Physics
at the address below.

Institute of Physics
Dirac House, Temple Back
Bristol BS1 6BE UK

Phone: 44 1 17 929 7481
Fax: 44 1 17 920 0979

techtracking@iop.org

Additional copies of this publication are available from:

Curran Associates, Inc.
57 Morehouse Lane
Red Hook, NY 12571 USA
Phone: 845-758-0400
Fax: 845-758-2633
Email: curran@proceedings.com
Web: www.proceedings.com

TABLE OF CONTENTS

Journal of Physics: Conference Series **752** (2016) 011001 doi:10.1088/1742-6596/752/1/011001

About the Conference:

Atoms, molecules, and light have been at the forefront of understanding quantum mechanics, since its discovery over 100 years ago. Since then the field has progressed from understanding the most fundamental aspects of how particles behave under quantum mechanics to controlling individual atoms for creating new technologies. While matter and light are from ordinary experience rather different, with today's control of coherent quantum phenomena many of the ideas freely cross their respective boundaries. Some of the topics that will be covered at the conference include, but are not limited to, cold atoms & cold molecules, ultrafast & precision spectroscopy, quantum manipulation & precision measurement, quantum computing & quantum communication, and quantum metrology. In this conference we aim to bring together the leading experts working in the frontiers of atomic, molecular, and optical systems.

Scientific Committee:

Journal of Physics: Conference Series **752** (2016) 011001 doi:10.1088/1742-6596/752/1/011001

Invited Speakers:

Jürgen Appel (Copenhagen, Denmark)
Luiz Davidovich (Universidade Federal Do Rio De Janeiro, Brazil)
Jonathan Dowling (LSU, USA)
Luming Duan (Michigan, USA)
Claude Fabre (Universite Pierre Et Marie Curie, France)
Elisabeth Giacobino (CNRS Laboratoire Kastler-Brossel, France)
Rudolf Grimm (Innsbruck, Austria)
Fedor Jelezko (Ulm, Germany)
Ruxin Li (SIOM,China)
Peter Van Loock (Mainz, Germany)
Longsheng Ma (ECNU, China)
Jian-Wei Pan (USTC, China)
Jean-Francois Roch (ENS Cachan, France)
Bao-Sen Shi (USTC, China)
Masahito Ueda (Tokyo, Japan)
Pengjun Wang (Shanxi, China)
Jörg Wrachtrup (Stuttgart, Germany)
Min Xiao (Arkansas, USA)
Lei Xu (Fudan, China)
Li You (Tsinghua, China)

Peer review statement

All papers published in this volume of *Journal of Physics: Conference Series* have been peer reviewed through processes administered by the proceedings Editors. Reviews were conducted by expert referees to the professional and scientific standards expected of a proceedings journal published by IOP Publishing.

Efficient creation of dipolar coupled nitrogen-vacancy spin qubits in diamond

I Jakobi[1], S A Momenzadeh[1], F Fávaro de Oliveira[1], J Michl[1], F Ziem[1], M Schreck[2], P Neumann[1], A Denisenko[1] and J Wrachtrup[1]

[1] 3. Physikalisches Institut, Unversität Stuttgart and Institute for Integrated Quantum Science and Technology IQST, Pfaffenwaldring 57, 70569 Stuttgart, Germany

[2] Experimentalphysik IV, Institut für Physik, Universität Augsburg, Universitätsstraße 1 Nord, 86159 Augsburg, Germany

Email: i.jakobi@physik.uni-stuttgart.de

Abstract. Coherently coupled pairs or multimers of nitrogen-vacancy defect electron spins in diamond have many promising applications especially in quantum information processing (QIP) but also in nanoscale sensing applications. Scalable registers of spin qubits are essential to the progress of QIP. Ion implantation is the only known technique able to produce defect pairs close enough to allow spin coupling via dipolar interaction. Although several competing methods have been proposed to increase the resulting resolution of ion implantation, the reliable creation of working registers is still to be demonstrated. The current limitation are residual radiation-induced defects, resulting in degraded qubit performance as trade-off for positioning accuracy. Here we present an optimized estimation of nanomask implantation parameters that are most likely to produce interacting qubits under standard conditions. We apply our findings to a well-established technique, namely masks written in electron-beam lithography, to create coupled defect pairs with a reasonable probability. Furthermore, we investigate the scaling behavior and necessary improvements to efficiently engineer interacting spin architectures.

1. Engineering dipolar coupled spin qubits in diamond

Engineering and scaling of coupled qubit systems or quantum registers lies at the core of quantum information processing as the expansion of quantum algorithms to useful applications depends on computational resources [1]. Solid state spin systems [2,3] have the prospect of chip-scale nanofabrication, where experience can be drawn from decades of development. Here, designed structures of coherently coupled spins, e.g. chains and arrays, are promising architectures for applications like quantum processors, simulators [4] or repeaters [5], and could find utilization in fields beyond quantum computation, for example as sensing arrays in quantum metrology [6,7]. Interestingly, in nanoscale metrology coupled spin dimers and trimers would boost the performance of such sensors by exploiting entanglement leading to useful improvements [8].

Single nitrogen-vacancy defect centers (NV) in diamond are known to be promising spin qubits and hybrid quantum registers [9]. A qubit defined on the $\mathrm{m_S} = 0, \pm 1$ sublevels of the electron spin ($\mathrm{S} = 1$) in the negative charge state can be controlled with static and dynamic magnetic fields, e.g. driving Rabi oscillations or inducing Larmor precessions, which gives rise to single qubit gate operations on a time scale of tens to hundreds of nanoseconds. In bulk samples long spin coherence times, typically on the order of tens to hundreds of μs and ranging up to ms at room-temperature [10], allow for many gate operations before the information stored on the qubit is lost. In addition, the spin can be initialized and readout optically, which allows to conveniently address single NV defects. The Hamiltonian describing the dynamics of a single NV spin can be found in the supplementary information.

The electron spin interacts with nuclear spins in the proximity of the defect. These can be for example paramagnetic ^{13}C (spin 1/2) in the diamond lattice or the intrinsic ^{14}N, or ^{15}N (spin 1 and ½, respectively). Nuclear spins can be controlled and readout via the NV electron spin through the hyperfine interaction [11,12] which allows single qubit gate operations as well as non-local gate operations between sets of nuclear spins and the central electron spin [9,13]. In addition, coherence times of nuclear spins can reach seconds at ambient conditions [14], making them an ideal storage resource for quantum information. Hence, qubits defined on the nuclear spins can expand a single NV to a versatile hybrid quantum register reaching sizes of up to four qubits [9]. However, such registers are not scalable as the coupling strength of suitable ^{13}C spins has to be larger than the inverse coherence lifetime [15] and the spectral resolution to address individual nuclear spins is limited by the longitudinal relaxation time of the central spin [16]. Ultimately, any scaling approach has to involve coupling multiple NV electron spin nodes between which quantum information can be shared.

There are several approaches to couple multiple NV electron spins. At cryogenic temperatures NV spins can be entangled with emitted photons, which are then processed with linear optics QIP schemes [17,18], or the electron spin can be coupled to the mechanical motion of nanoscale oscillators [19]. Here, we focus on an approach that is feasible at room temperature and does not rely on mediation but rather uses direct dipolar magnetic coupling [20-22], as illustrated in Figure 1 (a). Two electron spins A and B interact through their magnetic moments according to the interaction Hamiltonian

$$\hat{H}_{\mathrm{dip}} = \frac{\mu_0\, h^2\gamma^2}{4\,\pi\, r^3}\left(\vec{S}_{\mathrm{A}}\cdot\vec{S}_{\mathrm{B}} - 3\left(\vec{S}_{\mathrm{A}}\cdot\vec{r}\right)\left(\vec{S}_{\mathrm{B}}\cdot\vec{r}\right)\right), \qquad (1)$$

where μ_0 is the vacuum permeability, h is the Planck constant, $\gamma = 28.03$ GHz/T is the gyromagnetic ratio (divided by 2π) of the NV electron spin, $\vec{S_i} = R_i \cdot \vec{S_i'} = R_i \cdot \left(\hat{S}_{x,i}, \hat{S}_{y,i}, \hat{S}_{z,i}\right)^T \,|_{i=\mathrm{A,B}}$ are the dimensionless spin operators (expressed in a common frame through rotation matrices R_i), $\vec{r}$ is the unit vector connecting them and r is the distance. In the case of two NV spins the interaction can be reduced to an $\hat{S}_{z,\mathrm{A}}\,\hat{S}_{z,\mathrm{B}}$ coupling and approximated to

$$\hat{H}_{\mathrm{dip}} \approx h\nu_{\mathrm{dip}}\,\hat{S}_{z,\mathrm{A}}\,\hat{S}_{z,\mathrm{B}}, \qquad (2)$$

with the dipolar coupling strength

$$\nu_{\mathrm{dip}}(r, \phi_{\mathrm{AB}}, \phi_{\mathrm{Ar}}, \phi_{\mathrm{Br}}) = \frac{\mu_0\, h\,\gamma^2}{4\,\pi\, r^3}\left(\cos\phi_{\mathrm{AB}} - 3\cos\phi_{\mathrm{Ar}}\cos\phi_{\mathrm{Br}}\right). \qquad (3)$$

ϕ_{AB} is the angle between the two spin quantization axes, which correspond to the crystal axes of the respective defects, and $\phi_{\mathrm{Ar}}, \phi_{\mathrm{Br}}$ are the angles between each quantization axis and the connecting vector $\vec{r}$ (see supplementary information for a detailed derivation). All parameters of ν_{dip} depend on the relative position and orientation of the two spins which is thus a fixed property for each pair of NVs. With its inverse cubic scaling, dipolar coupling is a short-ranged interaction. For example, a pair of NV defects at a distance of about 33 nm and a median angular contribution couples with merely 1 kHz (see Figure 1 (d)) and even with ideal orientations of the dipoles the same coupling strength requires a distance of 47 nm.

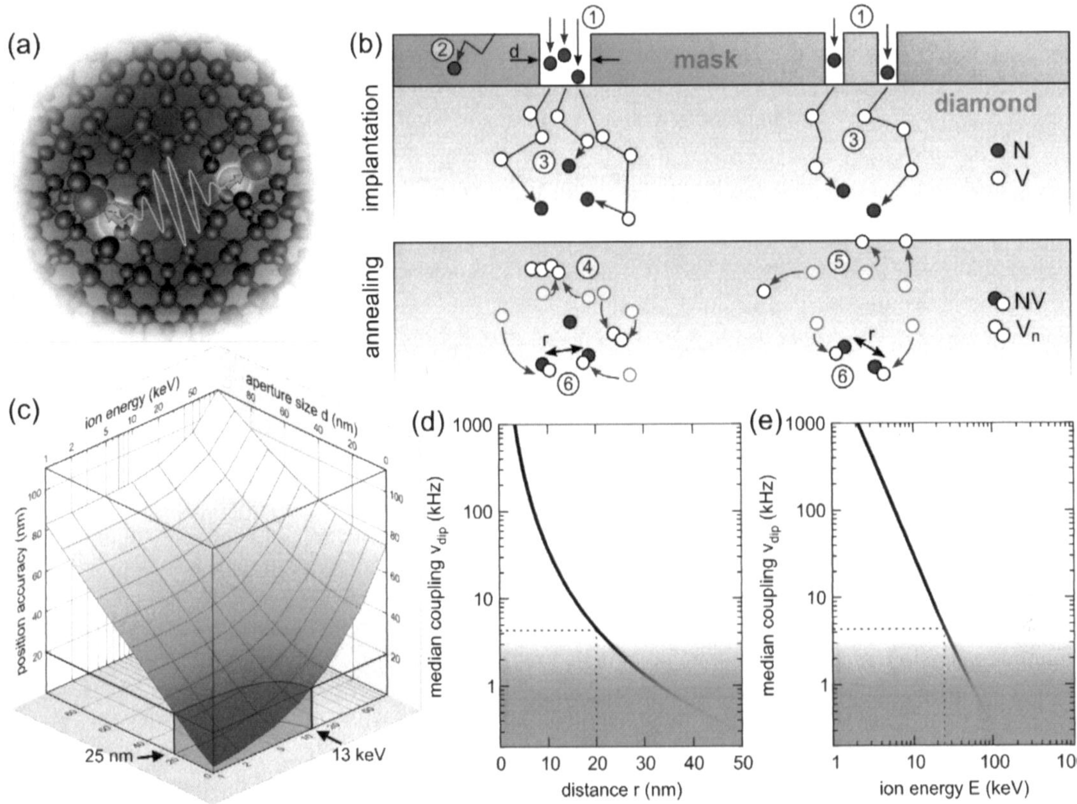

Figure 1 (a) Artwork of a coupling NV pair. **(b)** Schematic of nanomask implantation – Nitrogen ions (blue circles) are accelerated towards a diamond sample (1). A mask is able to block nitrogen radiation except for apertures of diameters d on the nano-scale (2). Ions passing through a single (left) or targeted through individual (right) apertures penetrate the diamond surface and scatter with lattice atoms (straggling), creating vacancies (white circles) in their tracks (3). The positioning accuracy depends on the aperture size and the straggling. During annealing vacancies become mobile while nitrogen atoms are stationary. Vacancies can form agglomerates (4), recombine at the surface or in the lattice (5) or combine with nitrogen atoms to form NV defects (6). **(c)** Positioning accuracy in terms of full width half maximum of the resulting ion distribution (ion resolution) depending on ion energy and aperture size d. The blue cut denotes the parameter range with an accuracy below 20 nm. **(d)** Coupling strength of two NV spins over their distance with a median angular contribution. With typical coherence times of implanted NVs ranging up to several hundreds of μs, couplings below a few kHz are unlikely to result in strong coupling (red gradient). This translates to an approximate threshold of 20 nm for the strong coupling regime. **(e)** Simulated median coupling strength of two NVs, implanted through the same point shaped aperture, depending on the ion energy.

Note that throughout the work we often use median instead of mean values as many simulated and experimental statistics depend on random sampling where the median is less prone to be influenced by outliers and better represents central tendencies.

In general, any pair of spins interacts through their magnetic moments, whilst in practice the distance is limited by the coherence times as the coupling can only have an effect if it acts on a shorter timescale than the relaxation. Typical detection schemes of the coupling involve preparing one spin (sensor) in a superposition state with a phase that is sensitive to the magnetic field originating from the second spin (emitter) which is prepared in a magnetic state $|m_S = \pm 1\rangle$. The detection is then limited by the transverse relaxation time T_2 of the sensor, as well as the longitudinal relaxation time T_1 of the emitter.

In order to exchange quantum information between the spins, i.e. create entangled states between two spins, both need to act as sensors at the same time. If only one spin fulfills the requirement they can only be classically correlated which we dub "weak coupling" in the following. The limitation of a quantum register consisting of two NV nodes can thus be expressed in the "strong coupling" condition

$$\nu_{\mathrm{dip}} > \frac{1}{\min(T_{2\mathrm{A}}, T_{2\mathrm{B}})}. \tag{4}$$

This is the primary limitation for the scaling of the NV quantum register via dipolar coupling. In essence, NVs need to be created at close distances to produce large couplings, while still having an unperturbed environment to retain long coherence times. Based on record coherence times of ms, the previously mentioned example of ν_{dip}=1 kHz hereby marks a threshold where pairs with distances larger than 50 nm will not fulfill the condition. With typical coherence times of tens to hundreds of μs, distances as small as 10-20 nm are required. Note that the effective coupling strength is four times larger if the spin projections $|\pm 1\rangle$ are utilized for non-local gates instead of projections $|0, +1\rangle$ or $|0, -1\rangle$ [20], which can in principle increase the distance by a factor of 1.6.

The established method to produce NVs with spatial precision on a scale of tens of nm is the implantation of nitrogen ions into diamond and subsequent conversion to NVs in an annealing process [23]. In fact, all previously reported pairs of NV defects exhibiting dipolar coupling [20-22] have been created using this process in different variations. Yet, despite advances in the precision only two strongly coupled pairs could be identified so far [21,22], showing the probabilistic nature and low success rate of the process.

The technique is based on accelerated nitrogen ions directed towards a diamond sample [24] with a kinetic energy E, as shown in Figure 1 (b). Upon impact on the surface they penetrate the lattice and scatter on the carbon atoms, creating vacancies along their tracks, until all kinetic energy is consumed and they come to a stop either as substitutional or interstitial defects. During the annealing step the former can be converted to NV defects by capturing mobile vacancies that typically originate from lattice damage in the ion tracks. The resolution, or spread of stopping positions, is influenced by the accuracy of the ion beam, i.e. the distribution of surface penetration points. It can be reduced by focusing techniques [25] or a variety of mask techniques where nanometer sized channels allow ions to pass a medium in defined locations while otherwise being blocked, e.g. nano-channels in mica [26], e-beam lithography on polymers [27,28], or pierced AFM tips [29]. While ions can be implanted through a single channel (Figure 1 (b) left) leaving exact impact positions to chance, the ideal approach to engineer structures of coupled NV defects is to position ions individually through multiple channels (Figure 1 (b) right). However, even with a perfect point source on the diamond surface the accuracy is fundamentally limited by the ion scattering in the lattice (straggling). Higher ion energies will result in larger straggling ranges, as shown in Figure 1 (c). For example ions with energies higher than about 10 keV have straggling beyond 20 nm (FWHM), resulting in median couplings smaller than a few kHz (Figure 1 (d)). Consequently, straggling favors low ion energies for the creation of dipolar coupled NV defects.

The annealing process serves to convert implanted nitrogen atoms to NV defects and heal lattice damages induced by the implantation. However, as vacancies can also bind together to form divacancy defects or larger agglomerates, not all damages are repaired which can introduce decoherence on proximal NV spins, impair the charge state or prevent the formation of NV defects in the first place. Ideally an ion should stop far away from most of the lattice damage and vacancies should have a low density in order to avoid their agglomeration. As this is only the case at high energies (~MeV), coherence times of implanted NV decrease rapidly for lower energies. For example, NV created by 5 keV implantation have coherence times on the order of a few μs [30] while NVs created in different processes at the same depth, for example by delta-doping, can in principle reach hundreds of μs [31]. The conversion yield of ions to NV defects also decreases for lower energies [32], as shown in Figure 2 (a). Hence, to create two NVs at close distances, either a high ion fluence has to be used, causing more implantation damage and shorter coherence times of the resulting qubits, or many sites have to be implanted with few ions resulting in a low probability to create two nearby NV defects. In addition, lower ion energies result in a shallow implantation, where proximity to the surface exposes NV defects to noisy environments of electron spins deposited in layers on the diamond [33,34]. Thus, the quality of qubits poses a challenge to the implantation coupled pairs at lower energies. This especially impairs the multi-channel approach which requires low ion energies to retain the accuracy.

The counteracting effects of yield, decoherence and straggling suggest that the creation of coupled NV defects using current processing methods has generally low success rates. At high energies, NV defects with long coherence times can be created with high probability, but straggling makes large dipolar coupling strengths unlikely. On the other hand, at low energies ions can be positioned reliably at close distances, but the coherence times of resulting NVs are likely insufficient to detect any couplings to their neighbors. The combination of these trade-offs suggests that the highest success rates to produce strongly coupled NVs are to be found in an intermediate energy range. In the following we quantify the interrelations in simulations for single-channel implantation, present experimental results of implanted coupled NV pairs and investigate the potential scaling through ion implantation.

2. Estimation of optimal implantation parameters for coupled defects

In order to estimate the optimal implantation parameters and the success rate of generating strongly coupled pairs of NV defects through a single channel, we conduct a Monte-Carlo study of the implantation process. The parameters of interest are the channel aperture (diameter d and area $A = d^2 \cdot \pi/4$), the ion energy E and the ion fluence $F = n_{N^+}/A$, denoting an amount of nitrogen ions n_{N^+} over an area A.

The ideal channel is a point source. In order to simulate a realistic case we choose a finite size. The highest success rates are expected for energies higher than a few keV. In these cases, the straggling is already on the order of tens of nm and aperture sizes of similar dimensions do not significantly reduce the overall position accuracy (see Figure 1 (c)). Hence, a diameter $d = 50$ nm that is technically feasible, e. g. via e-beam lithography in PMMA [27,28], is used.

The study is based on the strong coupling condition. We seek to find the ratio of implantation sites where at least two NV defects are created that fulfill this requirement. Thereby we need to model the amount of created NV defects, their coupling strength and their coherence times. In a single simulation run, the amount of ions n_{N^+} passing the channel is drawn from a Poissonian distribution with a mean value $\bar{n}_{N^+}$ depending on the fluence F and the aperture area A:

$$\bar{n}_{N^+} = F \cdot A = F \cdot \frac{\pi}{4} d^2. \tag{5}$$

Each ion receives a random lateral starting coordinate $\vec{r}_{\text{mask}}$ on the area of the aperture. Under the assumption of a homogeneous fluence throughout the aperture the whole area has a uniform probability of being hit by an ion. Afterwards the scattering of the ion in the diamond is simulated using stopping and range of ions in matter [35] where each ion receives a random displacement coordinate $\vec{r}_{\text{lattice}}$ based on its kinetic energy E. The stopping position is then obtained from the sum

$$\vec{r} = \vec{r}_{\text{mask}} + \vec{r}_{\text{lattice}} \,. \tag{6}$$

The amount of nitrogen ions that convert to NV defects during the annealing process can be modeled based on the empirical conversion yield data presented by Pezzagna, et al. [32] and shown in Figure 2 (a). The yield is assumed to be independent of the density of implanted ions, i.e. the ion fluence and each implanted ion has the same probability $y(E)$ to be converted into an NV defect that only depends on the energy E. The axis orientation for each NV defect is chosen at random from the set of four available crystal axes

$$\vec{a}_{\text{NV}} = \frac{1}{\sqrt{3}} \{ \begin{pmatrix} 1 \\ 1 \\ 1 \end{pmatrix}, \begin{pmatrix} 1 \\ -1 \\ -1 \end{pmatrix}, \begin{pmatrix} -1 \\ 1 \\ -1 \end{pmatrix}, \begin{pmatrix} -1 \\ -1 \\ 1 \end{pmatrix} \} \tag{7}$$

with equal probability. At this point all information to evaluate the dipolar coupling strength ν_{dip} between pairs of NV defects is available. The angular terms are obtained by the dot products

$$\cos \phi_{\text{AB}} = \vec{a}_{\text{A}} \cdot \vec{a}_{\text{B}} \,, \tag{8}$$

$$\cos\phi_{ir} = \frac{\vec{a}_i \cdot (\vec{r}_{\mathrm{A}} - \vec{r}_{\mathrm{B}})}{r} \quad |_{i=\mathrm{A,B}} \tag{9}$$

and the distance $r = |\vec{r}_{\mathrm{A}} - \vec{r}_{\mathrm{B}}|$.

The coherence times T_2 of implanted NV defects depend on many dynamics, e.g. spin-noise introduced by paramagnetic vacancy clusters from the implantation, substitutional and interstitial nitrogen defects, ^{13}C in the diamond lattice, surface spins and electric noise caused by unstable charge environments. As there is no complete theoretical framework that can reliably predict the coherence times of implanted NV defects we have to rely on empirical data to estimate the outcome depending on the ion energy E and the fluence F. The used data was collected throughout the characterization of several samples implanted both through PMMA nano-apertures and without mask, and recorded as NV ensembles, at different energies and fluences. An observational model of the median coherence times using a logistic growth over the ion energy and an exponential decay over the ion fluence is fitted to the data and is shown in Figure 2 (b). Further details can be found in the supplementary material. For each NV defect taken in the Monte Carlo simulation a coherence time is drawn from an exponential distribution with the appropriate median value. Finally, the strong coupling condition of defect pairs can be evaluated.

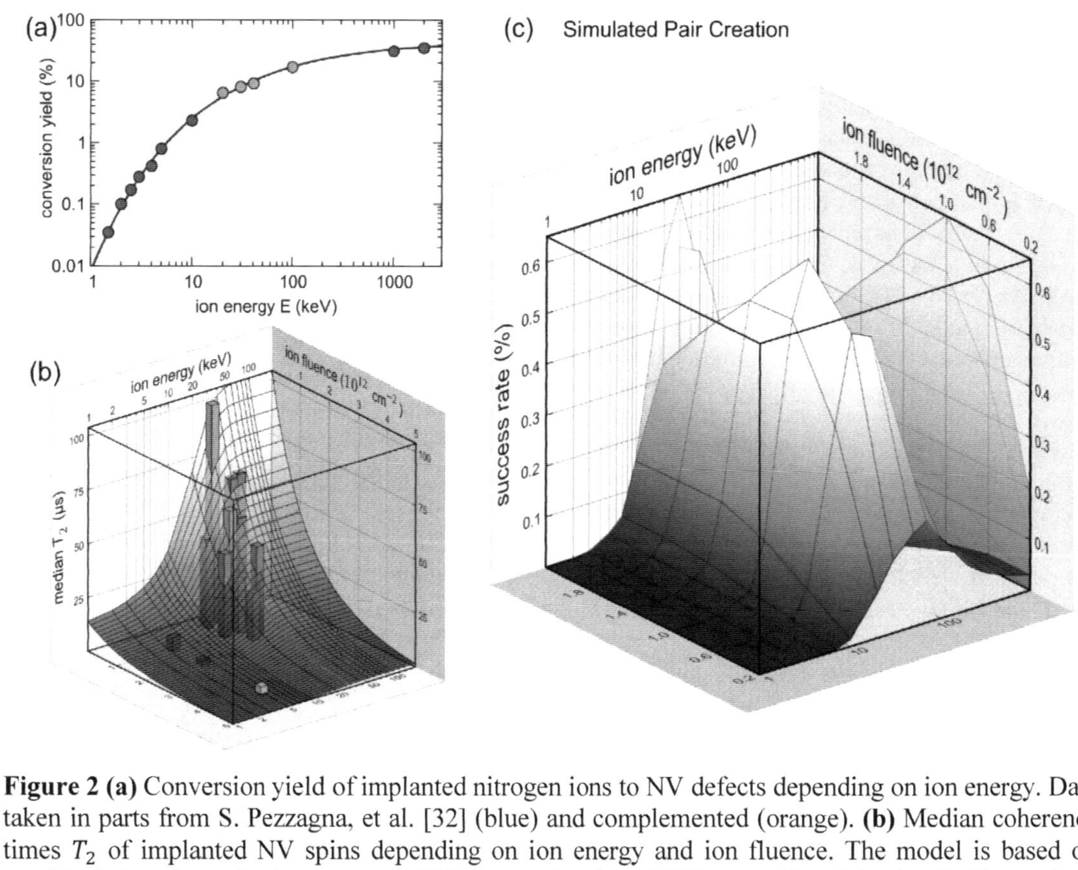

Figure 2 (a) Conversion yield of implanted nitrogen ions to NV defects depending on ion energy. Data taken in parts from S. Pezzagna, et al. [32] (blue) and complemented (orange). **(b)** Median coherence times T_2 of implanted NV spins depending on ion energy and ion fluence. The model is based on empirical data from single nanomask NVs (green bars) and NV ensembles (blue bars). The optimal simulated parameters in subfigure (c) have a median coherence times of 41 μs (red guides). **(c)** Simulated success rate of implanting pairs of NV defects through a 50 nm aperture where the dipolar coupling strength is larger than the decoherence rates of both spins. A maximum probability of 0.6% is reached at an energy of 30 keV and a fluence of 10^{12} cm^{-2} corresponding to an average of 20 implanted ions within one site.

The results for several ion energies and fluences, each sampled over one million runs, is shown in Figure 2 (c). As expected, the highest success rates of implanting strongly coupled pairs of NV defects are found in an intermediate range of 20-50 keV. The maximum efficiency of 0.6 % is found at $E = 30$ keV and $F = 10^{12}$ cm^{-2}, corresponding to around 20 ions per site with an aperture area of $\pi/4 \cdot 50^2$ nm^2 and an average of 1.6 created NV per site with an expected yield of 8 % for an energy of 30 keV. Lower fluences will reduce the probability for two NV defects to be created and higher fluences result in shorter coherence times. The energy range below 5 keV that is suitable for a multi-channel approach, has success rates on the order of 0.025 %. Even when the overall coupling strengths are increased by simulating a point shaped aperture the highest success rates at 5 keV merely reach values of 0.09 %.

3. Experimental realization of coupled defects

With simulated probabilities below 1% the process is far from deterministic. Post-selecting sites from a set of several hundreds or thousands sites is, however, realistic. We therefore apply the simulated implantation parameters and prepare two identical [100]-oriented, single-crystal, CVD-grown diamond samples with natural abundance of ^{13}C (1.1%). To form the implantation mask, we employed e-beam lithography technique using PMMA positive resist. An acceleration voltage of 30 kV and an aperture size of 10 μm were used to expose approximately 5,000 sites per sample with a diameter of 50 nm in the spin-coated PMMA (50K 9%) with the thickness of approximately 400 nm. Subsequently, $^{14}N^+$ ions are implanted with the calculated fluence of 10^{12} cm^{-2} and energy of 30 keV with 0° incident angle (see Figure 3 (a)). After removing the PMMA masks the samples are annealed for two hours at 950°C in a 10^{-7} mbar vacuum. The carbon lattice is terminated with oxygen by boiling the diamond samples in acid in order to have the implanted NV defects in the negative charge state.

Figure 3 (b) shows an exemplary confocal fluorescence scan of NV sites implanted through the grid of nano-apertures. The amount of created NV defects per implantation site is assessed by their fluorescence intensity. We find a Poissonian distribution with an average of 1.6 NV defects per site (1.69 on sample A, 1.54 on sample B, see Figure 3 (c)), proving that the experimental parameters match the simulation. Throughout the characterization no significant difference between the two samples was found.

Each implantation site that is bright enough to host two or more NV defects is investigated for dipolar coupling. The spin resonances of parallel oriented NV spins overlap and thus individual addressing in double electron-electron resonance (DEER) experiments is challenging. Hence, one quarter of potentially coupling pairs are not identified. The coherence times for each addressable NV are determined by conducting electron spin echo envelope modulation (ESEEM) measurements in an axially aligned magnetic field of 5 mT. A histogram of measured coherence times is shown in Figure 3 (d), where an exponential distribution with a median T_2 of 33 μs (41 μs expected) can be seen. If a measured coherence time surpasses a threshold of 20 μs, corresponding to a decoherence rate of 50 kHz, the NVs are tested for dipolar coupling in a DEER experiment. For a detailed description of ESEEM and DEER experiments see the supplementary information.

For the characterization of a large amount of implantation sites an automated routine is employed. This has the advantage that implantation sites can be consistently investigated within an average measurement time of five minutes and without subjective human bias. On the other hand we expect the routine to produce false negatives without any indication of the failure rate. For example, the evolution times in DEER experiments were only measured for up to 100 μs, setting a lower bound of around 10 kHz for the detectable coupling. Positive results are always subject to validation in separate experiments.

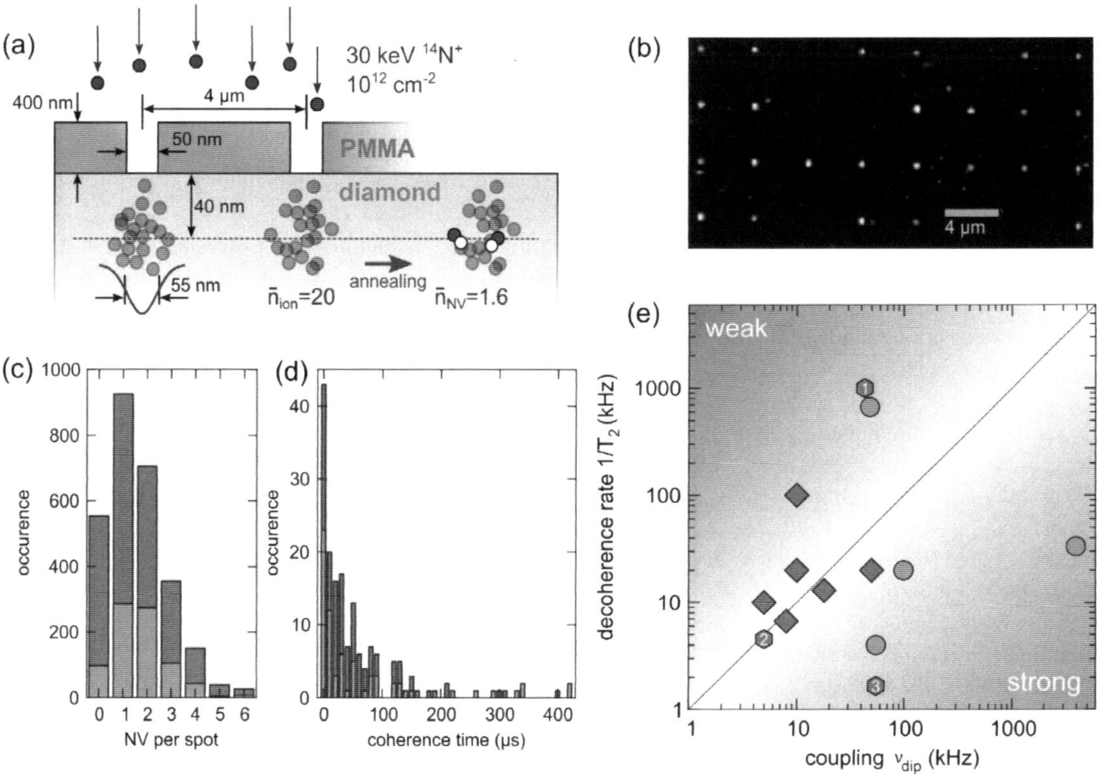

Figure 3 (a) Schematics of the implantation of $^{14}N^+$ ions through 50 nm apertures in a 400 nm PMMA mask. The spacing between holes is 4 μm. The simulated optimal parameters are used. This results in 20 implanted ions and two created NV defects on average, a depth of 40 nm and a resolution of 55 nm (FWHM). A total of 10,000 sites were implanted throughout two samples. **(b)** Confocal microscopy image of an implanted region with a grid of NV sites. **(c)** Histogram of amount of implanted NV defects per site counted in 818 and 1941 sites on sample A (orange) and B (blue), respectively. The average amount is 1.58 NVs per site. **(d)** Histogram of T_2 coherence times (ESEEM) of implanted NV recorded on 77 and 96 sites on sample A (orange) and B (blue), respectively. The median coherence time is 33 μs. **(e)** Characteristics of identified coupled NV pairs. For each found pair, the limiting decoherence rate is plotted over the dipolar coupling strength. The diagonal line marks the threshold of the strong coupling condition. Weakly and strongly coupled pairs appear in the top-left and bottom-right regions, respectively. Pairs found in sample A are marked as orange circles, sample B as blue diamonds. Previously reported pairs from P. Neumann, et al. (1), F. Dolde, et al. (2) and T. Yamamoto, et al. (3) are marked as grey hexagons.

The search was conducted over 6,000 implantation sites, amassing a total measurement time of about three weeks. Throughout both samples a total of six strongly coupled and four weakly coupled pairs were identified. The coupling strengths and limiting decoherence rates are shown in Figure 3 (e) and individual measurements are presented in the supplementary information. Most identified coupling strengths appear in the range of few tens of kHz. Lower couplings are hardly detected by the search routine. Only two couplings with 5 and 8 kHz could be identified in this range. The strongest recorded coupling strength is 3.9 MHz corresponding to maximum distance of 3 nm for an ideal angular constellation. The overall success rate of 0.1 % is within the same order of magnitude as the simulated 0.6 %. The actual amount of created pairs is expected to be larger because of omitted parallel NVs and false negatives from the search routine. No trimers or larger networks of coupled NV spins could be identified.

The presented experimental single-hole implantation has room for improvement. For example, the aperture size of 50 nm can be shrunk to smaller dimensions, decreasing the size of the positioning distribution from 55 nm (FWHM) by up to 45 % to 30 nm for a 10 nm hole. According to simulations this would increase the chance to create strongly coupled NV pairs by a factor of about 1.8. The coherence times of implanted NVs can also be improved, for example by using isotopically enriched ^{12}C samples. However, hyperfine coupled ^{13}C nuclear spin qubits might be an important resource for the scaling of the NV quantum register. Furthermore, it is doubtful that the ^{13}C spin bath is limiting the NV coherence times for the desired low energy implantation range. The cause here are rather implantation damages and surface spins. Surface effects can be mitigated by growing an additional CVD layer after the implantation[30].

4. Further scaling

Although the statistic sample sizes are relatively small, the experiments show comparable results to the simulation. With this affirmation we investigate the probabilities for larger coupled networks using the same method. We define a network of size *n* as any amount of NVs where *n* defects are interconnected by strong coupling without interruption. More details can be found in the supplementary information.

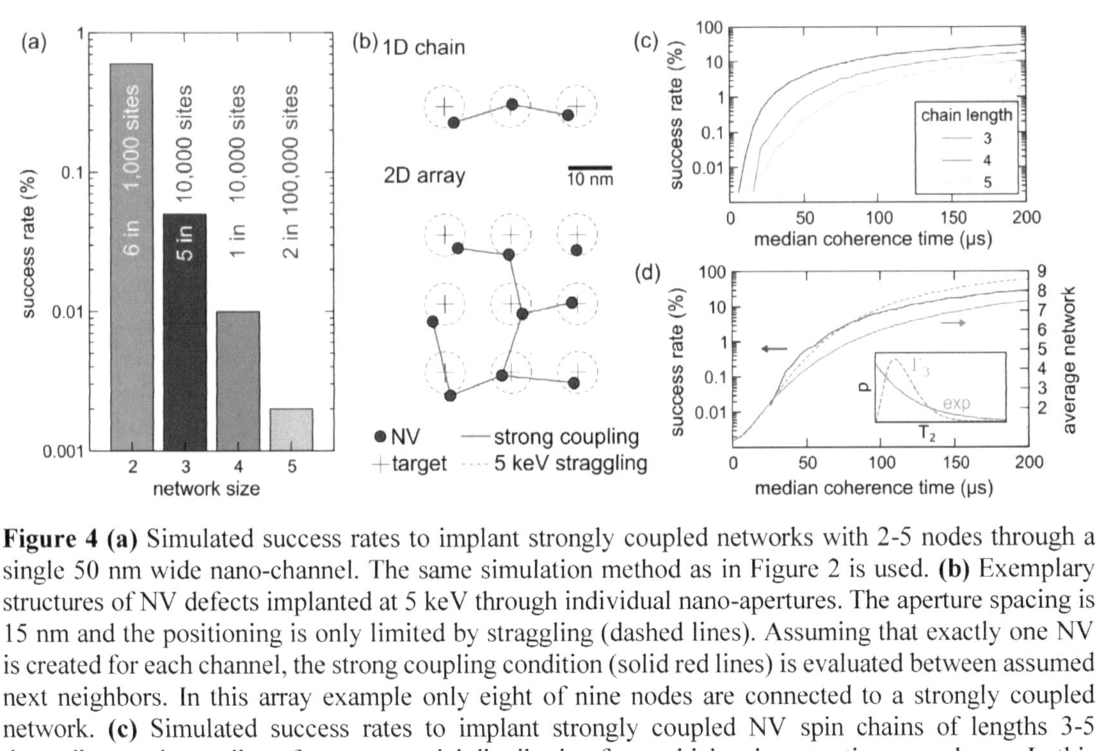

Figure 4 (a) Simulated success rates to implant strongly coupled networks with 2-5 nodes through a single 50 nm wide nano-channel. The same simulation method as in Figure 2 is used. **(b)** Exemplary structures of NV defects implanted at 5 keV through individual nano-apertures. The aperture spacing is 15 nm and the positioning is only limited by straggling (dashed lines). Assuming that exactly one NV is created for each channel, the strong coupling condition (solid red lines) is evaluated between assumed next neighbors. In this array example only eight of nine nodes are connected to a strongly coupled network. **(c)** Simulated success rates to implant strongly coupled NV spin chains of lengths 3-5 depending on the median of an exponential distribution from which coherence times are drawn. In this approach coherence times of tens to 100 μs are required for realistic success rates. **(d)** Simulated success rates (blue, left y-axis) to implant a strongly coupled 3x3 NV spin array depending on the same coherence time distributions and average resulting network size (red, right y-axis). Coherence times are drawn from exponential distributions (solid) and gamma distributions with shape-factor 3. The inset shows exemplary distributions.

The success rates to create triples, quadruples and quintuples are shown in Figure 4 (a), each for their respective optimal implantation parameters. For each additional node the probability decreases by about one order of magnitude. Extrapolating from the experiment a triple might require the investigation of an order of 10,000 implantation sites and months of measurement time to be found, whereas searching for any larger network appears unfeasible. One would expect that as the density of NV defects increases, the average distance decreases, favoring the creation of larger networks. However, in order to create the necessary densities fluences have to be increased and the decoherence rates outscale the coupling. In order to achieve realistic probabilities to scale the NV quantum registers either the coupling strengths at high energies or coherence times at low energies have to be increased. Straggling is inherent to implantation and unlikely to be circumventable, whereas coherence times of NV defects implanted at low energies may be improved through optimized processing methods during implantation and annealing[30,36,37].

Revisiting the multi-channel approach we conduct a theoretical study to estimate the necessary improvement in coherence times to make engineered NV spin chains and arrays feasible. In the ideal case exactly one NV is created for each nano-channel and the accuracy only depends on the fundamental straggling. In order to reach adequate coupling strengths of few kHz to tens of kHz, the distance between NVs should be within 20 nm. For this example an implantation site spacing of 15 nm and an ion energy of 5 keV used, which is technically feasible, for example with pierced AFM tips, or within reach with lithographically written nano-apertures. An exemplary one dimensional chain and a two-dimensional array of NV defects are shown in Figure 4 (b). In a Monte-Carlo simulation NV positions and axis orientations are distributed, coherence times are drawn from exponential distributions and the couplings between next neighbor NV nodes are calculated.

The spin chains are investigated for broken links. Figure 4 (c) shows the success rate depending on the median coherence time for different lengths. For this example coherence times have to reach 50 to 100 μs in order to achieve single percentage success rates. Coherence times of this magnitude are within reasonable reach for the chosen implantation energy [30], yet have still to be realized for fluences that could reliably produce an NV defect in each implantation site. A spin chain has also the drawback that each link has to be strongly coupled, which is reflected in the decreasing success rates of longer chains. In two dimensions each node can have several links and unconnected or broken nodes can still be tolerated.

The array is therefore investigated for the largest network of strongly coupled NV defects and compared to the success rate to achieve a network with all nine nodes connected, as shown in Figure 4 (d). While the success rate of the array scales similarly to the chains, requiring again at least 50 to 100 μs in order achieve single digit percentages, the average largest network size already reaches values of three to six nodes in the same range. Here the main reason for broken nodes or links are short coherence times that frequently occur. Assuming that with increased median values particularly short values can be avoided at the same time, we also simulate gamma distributions with a shape-factor of 3 (Γ_3) that have a gradual onset. At low median values there is no difference between the distributions. Around 30-50 μs, however, the Γ_3-distribution gains a significant advantage, which is the range where decoherence rates start scaling below the couplings. Hence not only overall coherence times of implanted NV defects have to be increased to a range that is already achieved with different generation approaches [30,31], but in particular short coherence times have to be avoided. With such improvements the probabilistic scaling of the NV quantum register appears to be feasible.

5. Conclusions

In conclusion we have devised a simulation framework that estimates the outcome of the implantation of coupled NV defects. Based on this we experimentally realized six strongly coupled pairs of NV defects, with coupling strengths of up to 4 MHz. While the success rate is far from ideal, our method signifies a substantial increase compared to previously reported techniques. Our recipe can be reproduced and may facilitate further research into dynamics of coupled solid-state spins, for example spatially correlated spectroscopy, entanglement protocols and quantum information processing in general.

While the presented Monte-Carlo simulations produced satisfactory results, they still rely on rudimentary models based on empirical data. The scattering of nitrogen ions in diamond is well understood. However, the dynamics of vacancies during the annealing process and the resulting conversion yield and coherence times of implanted NV defects lack comprehensive models. Progress in this field has the potential to improve the prediction results. Especially for future engineering of complex spin-chains and arrays, precise simulations may become invaluable tools to sieve through a vast parameter range in search for optimal implantation conditions.

Our work also shows that the creation process with contemporary methods is mostly probabilistic and generally showing success rates below 1% for pairs. While the precise positioning of NV defects certainly is important, especially for tailored constellations of coupled spins, the limiting factor for the scaling are the coherence times of implanted NV defects. Here processing methods need to increase the conversion yield and coherence times, for example through epitaxial overgrowth, in order to harness the benefits of already established precise ion positioning methods at low implantation energies.

References

[1] Nielsen M A and Chuang I L 2000 *Quantum Computation and Quantum Information* (Cambridge: Cambridge University Press)

[2] Shulman M D, Dial O E, Harvey S P, Bluhm H, Umansky V and Yacoby A 2012 *Science* **336** 202-205

[3] Muhonen J T, Dehollain J P, Laucht A, Hudson F E, Kalra R, Sekiguchi T, Itoh K M, Jamieson D N, McCallum J C, Dzurak A S, *et al.* 2014 *Nat. Nanotechnol.* **9** 986-991

[4] Cai J M, Retzker A, Jelezko F and Plenio M B 2013 *Nat. Phys.* **9** 168-173

[5] Jiang L, Taylor J M, Nemoto K, Munro W J, Van Meter R and Lukin M D 2009 *Phys. Rev. A.* **79** 032325

[6] Steinert S, Dolde F, Neumann P, Aird A, Naydenov B, Balasubramanian G, Jelezko F and Wrachtrup J 2010 *Rev. Sci. Instrum.* **81** 043705

[7] Simpson D A, Tetienne J P, McCoey J M, Ganesan K, Hall L T, Petrou S, Scholten R E and Hollenberg L C L 2016 *Sci. Rep.* **6** 22797

[8] Chin A W, Huelga S F and Plenio M B 2012 *Phys. Rev. Lett.* **109** 233601

[9] Waldherr G, Wang Y, Zaiser S, Jamali M, Schulte-Herbruggen T, Abe H, Ohshima T, Isoya J, Du J F, Neumann P, *et al.* 2014 *Nature* **506** 204-209

[10] Balasubramanian G, Neumann P, Twitchen D, Markham M, Kolesov R, Mizuochi N, Isoya J, Achard J, Beck J, Tissler J, et al. 2009 *Nat. Mater.* **8** 383-387

[11] Jelezko F, Gaebel T, Popa I, Domhan M, Gruber A and Wrachtrup J 2004 *Phys. Rev. Lett.* **93** 130501

[12] Childress L, Dutt M V G, Taylor J M, Zibrov A S, Jelezko F, Wrachtrup J, Hemmer P R and Lukin M D 2006 *Science* **314** 281-285

[13] Neumann P, Mizuochi N, Rempp F, Hemmer P, Watanabe H, Yamasaki S, Jacques V, Gaebel T, Jelezko F and Wrachtrup, J. 2008 *Science* **320** 1326-1329

[14] Maurer P C, Kucsko G, Latta C, Jiang L, Yao N Y, Bennett S D, Pastawski F, Hunger D, Chisholm N, Markham M, *et al.* 2012 *Science* **336** 1283-1286

[15] Zaiser S, Rendler T, Jakobi I, Wolf T, Lee S Y, Wagner S, Bergholm V, Schulte-Herbruggen T, Neumann P and Wrachtrup J 2016 *Nat. Commun.* **7** 12279

[16] van der Sar T, Wang Z H, Blok M S, Bernien H, Taminiau T H, Toyli D M, Lidar D A, Awschalom D D, Hanson R and Dobrovitski V V 2012 *Nature* **484** 82-86

[17] Bernien H, Hensen B, Pfaff W, Koolstra G, Blok M S, Robledo L, Taminiau T H, Markham M, Twitchen D J, Childress L, et al. 2013 *Nature* **497** 86-90

[18] Knill E, Laflamme R and Milburn G J 2001 *Nature* **409** 46-52

[19] Rabl P, Cappellaro P, Dutt M V G, Jiang L, Maze J R and Lukin M D 2009 *Phys. Rev. B.* **79** 041302

[20] Neumann P, Kolesov R, Naydenov B, Beck J, Rempp F, Steiner M, Jacques V, Balasubramanian G, Markham M, Twitchen D J, *et al.* 2010 *Nat. Phys.* **6** 249-253

[21] Dolde F, Jakobi I, Naydenov B, Zhao N, Pezzagna S, Trautmann C, Meijer J, Neumann P, Jelezko F and Wrachtrup J 2013 *Nat. Phys.* **9** 139-143

[22] Yamamoto T, Muller C, McGuinness L P, Teraji T, Naydenov B, Onoda S, Ohshima. T, Wrachtrup J, Jelezko F and Isoya J 2013 *Phys. Rev. B* **88** 201201
[23] Pezzagna S, Rogalla D, Wildanger D, Meijer J and Zaitsev A 2011 *New J. Phys.* **13** 035024
[24] Meijer J, Burchard B, Domhan M, Wittmann C, Gaebel T, Popa I, Jelezko F and Wrachtrup J 2005 *Appl. Phys. Lett.* **87** 261909
[25] Schnitzler W, Linke N M, Fickler R, Meijer J, Schmidt-Kaler F and Singer K 2009 *Phys. Rev. Lett.* **102** 070501
[26] Pezzagna S, Rogalla D, Becker HW, Jakobi I, Dolde F, Naydenov B, Wrachtrup J, Jelezko F, Trautmann C and Meijer J 2011 *Phys. Status Solidi A* **208** 2017-2022
[27] Toyli D M, Weis C D, Fuchs G D, Schenkel T and Awschalom D D 2010 *Nano Lett.* **10** 3168-3172
[28] Scarabelli D, Trusheim T, Gaathon O, Englund D and Wind S J 2016 *Nano Lett. Article ASAP*
[29] Pezzagna S, Wildanger D, Mazarov P, Wieck A D, Sarov Y, Rangeow I, Naydenov B, Jelezko F, Hell S W and Meijer J 2010 *Small* **6** 2117-2121
[30] Staudacher T, Ziem F, Haussler L, Stohr R, Steinert S, Reinhard F, Scharpf J, Denisenko A and Wrachtrup J 2012 *Appl. Phys. Lett.* **101** 212401
[31] Ohno K, Heremans F J, Bassett L C, Myers B A, Toyli D M, Jayich A C B, Palmstrom C J and Awschalom D D 2012 *Appl. Phys. Lett.* **101** 082413
[32] Pezzagna S, Naydenov B, Jelezko F, Wrachtrup J and Meijer J 2010 *New J. Phys.* **12** 065017
[33] Tisler J, Balasubramanian G, Naydenov B, Kolesov R, Grotz B, Reuter R, Boudou J P, Curmi P A, Sennour M, Thorel A, *et al.* 2009 *ACS Nano* **3** 1959-1965
[34] Ofori-Okai B K, Pezzagna S, Chang K, Loretz M, Schirhagl R, Tao Y, Moores B A, Groot-Berning K, Meijer J and Degen C L 2012 *Phys. Rev. B* **86** 081406
[35] Ziegler J F, Ziegler M D and Biersack J P 2010 *Nucl. Instrm. Meth. B* **268** 1818-1823
[36] de Oliveira F F, Momenzadeh S A, Wang Y, Konuma M, Markham M, Edmonds A M, Denisenko A and Wrachtrup J 2015 *Appl. Phys. Lett.* **107** 073107
[37] de Oliveira F F, Momenzadeh S A, Antonov D, Scharpf J, Osterkamp C, Naydenov B, Jelezko F, Denisenko A and Wrachtrup J 2016 *Nano Lett.* **16** 2228-2233

Acknowledgments

We acknowledge financial support by the DFG (SFB-TR 21, SPP1601 and FOR1493), the EU (SIQS), the DIADEMS consortium and the MPG. Furthermore, we thank Wolfgang Brückner and Olga Lik for their support with the implantation.

Multiple Period States of the Superfluid Fermi Gas in an Optical Lattice

Gentaro Watanabe[1,2], Sukjin Yoon[3], Franco Dalfovo[4] and Takashi Nakatsukasa[5,6]

[1] Department of Physics and Zhejiang Institute of Modern Physics, Zhejiang University, Hangzhou, Zhejiang 310027, China
[2] Center for Theoretical Physics of Complex Systems, Institute for Basic Science (IBS), Daejeon 34051, Korea
[3] School of Computational Sciences, Korea Institute for Advanced Study (KIAS), Seoul 02455, Korea
[4] INO-CNR BEC Center and Department of Physics, University of Trento, 38123 Povo, Italy
[5] Center for Computational Sciences, University of Tsukuba, Tsukuba 305-8577, Japan
[6] RIKEN Nishina Center, Wako, Saitama 351-0198, Japan

E-mail: gentaro@zju.edu.cn

Abstract. We study multiple period states (i.e., states whose period is a multiple of the lattice constant) of a two-component unpolarized superfluid Fermi gas in an optical lattice along the crossover between the Bardeen-Cooper-Schrieffer (BCS) and Bose-Einstein condensate (BEC) states. By solving Bogoliubov-de Gennes equations for a superfluid flow with finite quasimomentum, we find that, in the BCS side of the crossover, the multiple period states can be energetically favorable compared to the normal Bloch states and their survival time against dynamical instability drastically increases, suggesting that these states can be accessible in current experiments. This is in sharp contrast to the situation in BECs.

1. Introduction: nonlinear phenomena in superfluid Fermi gases in an optical lattice

Interplay between the nonlinearity due to the emergence of the superfluid order parameter and the periodicity of the lattice is very intriguing because these two are essential elements in condensed matter. Ultracold atomic gases in optical lattices enable us to study the subtle interplay of these effects deeply and directly (see, e.g., [1] for review) because of their high controllability of both the lattice geometry and the interatomic interaction (characterized by the s-wave scattering length a_s) (see, e.g., [2, 3]). Especially, by changing the interatomic interaction in superfluid Fermi gases using a Feshbach resonance, one can go along the crossover from the weakly coupled Bardeen-Cooper-Schrieffer (BCS) state to the Bose-Einstein condensate (BEC) state of tightly bound bosonic dimers [4, 5, 6, 7, 8], which allows us to study Bose and Fermi superfluids from a unified perspective.

Emergence of multiple period states, namely a class of stationary states whose period does not coincide with that of the external potential, but is a multiple of it, is a typical nonlinear phenomenon. For BECs in a periodic potential, it was found that multiple period states appear due to nonlinearity of the interaction term of the Gross-Pitaevskii (GP) equation [9]. However, these multiple period states in BECs are energetically unfavorable compared to the normal Bloch states whose period is equal to the lattice constant, and the lowest multiple period states are dynamically unstable [9].

Nonlinear phenomena in Fermi superfluids in a periodic potential [10, 11, 12] can be more important compared to those in Bose superfluids because of the wide implications for various systems in condensed matter physics and nuclear physics such as superconducting electrons in solids and superfluid neutrons in neutron stars (e.g., [13, 14, 15]). However, unlike the case of Bose gases, the study of nonlinear phenomena of superfluid Fermi gases is at a very infant stage (see, e.g., [16, 17]) and little has been studied about multiple period states in superfluid Fermi gases. Therefore, even a fundamental problem whether multiple period states exist along the BCS-BEC crossover was still open.

Under such circumstances, we have studied multiple period states in superfluid Fermi gases in [18]. In the present article, we summarize main results of this work (see also [1]).

2. Setup of the problem and formalism

We consider ultracold superfluid Fermi gases in the BCS-BEC crossover in a three-dimensional (3D) system flowing through a 1D optical lattice,

$$V_{\text{ext}}(\mathbf{r}) = V_{\text{ext}}(z) \equiv sE_R \sin^2 q_{\text{B}} z, \tag{1}$$

where s is the dimensionless parameter characterizing the lattice strength, $E_R \equiv \hbar q_B^2/2m$ is the recoil energy, $q_B \equiv \pi/d$ is the Bragg wave vector, and d is the lattice constant. We assume that the system is uniform in the transverse directions. We approach this problem by numerical simulations based on the Bogoliubov-de Gennes (BdG) equations [19, 8] for a supercell containing a multiple number of unit cells,

$$\begin{pmatrix} H'(\mathbf{r}) & \Delta(\mathbf{r}) \\ \Delta^*(\mathbf{r}) & -H'(\mathbf{r}) \end{pmatrix} \begin{pmatrix} u_i(\mathbf{r}) \\ v_i(\mathbf{r}) \end{pmatrix} = \epsilon_i \begin{pmatrix} u_i(\mathbf{r}) \\ v_i(\mathbf{r}) \end{pmatrix} \tag{2}$$

with $H'(\mathbf{r}) = -\frac{\hbar}{2m}\nabla^2 + V(\mathbf{r}) - \mu$ and μ being the chemical potential. Here, $v_i(\mathbf{r})$ and $u_i(\mathbf{r})$ are the quasiparticle amplitudes, associated with the probability of occupation and unoccupation of a paired state denoted by an index i, and ϵ_i is the corresponding eigenenergy. The quasiparticle amplitudes $v_i(\mathbf{r})$ and $u_i(\mathbf{r})$ satisfy the normalization condition, $\int d\mathbf{r}\,[u_i^*(\mathbf{r})u_j(\mathbf{r}) + v_i^*(\mathbf{r})v_j(\mathbf{r})] = \delta_{i,j}$. Δ is the order parameter (or the pairing field), which reduces to the pairing gap in the single quasiparticle spectrum in the region of $\mu > 0$ for the uniform system. The pairing field $\Delta(\mathbf{r})$ and the chemical potential μ in equation (2) are self-consistently determined from the gap equation,

$$\Delta(\mathbf{r}) = -g\sum_i u_i(\mathbf{r})v_i^*(\mathbf{r}) \tag{3}$$

with g being the coupling constant for the s-wave contact interaction which needs to be renormalized [20, 21, 22], and the average number density

$$n_0 = \frac{N}{\mathcal{V}} = \frac{1}{\mathcal{V}}\int n(\mathbf{r})\,d\mathbf{r} = \frac{2}{\mathcal{V}}\sum_i \int |v_i(\mathbf{r})|^2 d\mathbf{r}\,. \tag{4}$$

In the following, E_F and k_F denote the Fermi energy and wavenumber of a uniform free Fermi gas with density n_0: $E_F = \hbar^2 k_F^2/(2m)$ and $k_F = (3\pi^2 n_0)^{1/3}$. Since Δ depends on $\{u_i\}$ and $\{v_i\}$, the BdG equations (2) are nonlinear for nonzero g.

In this formalism, a stationary motion of the superfluid in the z direction, relative to the infinite periodic potential at rest, is described by solutions of equation (2) with quasimomentum

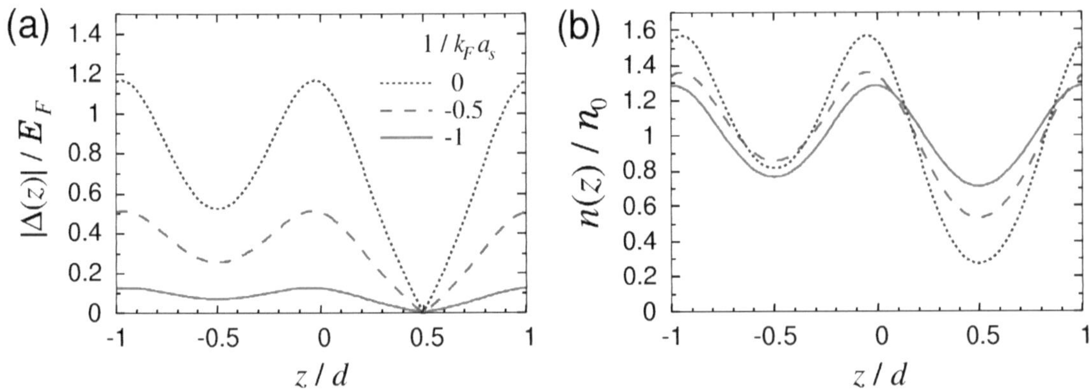

Figure 1. Profiles of (a) the magnitude of the pairing field $|\Delta(z)|$ and (b) the density $n(z)$ of the lowest period-doubled states at their first Brillouin zone edge $P = P_{\rm edge}/2 = \hbar q_B/4$ for various values of $1/k_F a_s$. Other parameters are $s = 1$ and $E_F/E_R = 0.25$. In the deep BCS side ($1/k_F a_s = -1$), unlike $|\Delta(z)|$, there is no large difference in $n(z)$ between the regions of $-1 < z/d \leq 0$ and $0 < z/d \leq 1$. However, this difference becomes more significant with increasing $1/k_F a_s$. This figure is taken from [18].

P per atom (not per pair), or the corresponding wave vector $Q = P/\hbar$, such that the quasiparticle amplitudes can be written in the Bloch form as $u_i(\mathbf{r}) = \tilde{u}_i(z)e^{iQz}e^{i\mathbf{k}\cdot\mathbf{r}}$ and $v_i(\mathbf{r}) = \tilde{v}_i(z)e^{-iQz}e^{i\mathbf{k}\cdot\mathbf{r}}$ leading to the pairing field as $\Delta(\mathbf{r}) = e^{i2Qz}\tilde{\Delta}(z)$. Here, $\tilde{\Delta}(z)$, $\tilde{u}_i(z)$, and $\tilde{v}_i(z)$ are periodic functions with period ν times d, with $\nu \in \{1, 2, 3, \cdots\}$, and the wave vector k_z of the quasiparticle lies in the first Brillouin zone for a supercell (a cell containing several primitive cells) with period ν, i.e., $|k_z| \leq q_B/\nu$.

3. Results

In figure 1, we show the magnitude of the pairing field $|\Delta(z)|$ and the number density $n(z)$ of the lowest period-doubled states (i.e., period $2d$) at their first Brillouin zone edge $P = P_{\rm edge}/2 = \hbar q_B/4$ ($P_{\rm edge} \equiv \hbar q_B/2$ denotes the quasimomentum at the first Brillouin zone edge of the normal Bloch states with period d). At this value of $P = P_{\rm edge}/2$, reflecting the fact that the current is zero, $|\Delta(z)|$ of the period-2 states has a node. The period-2 nature and the difference between the regions of $-1 < z/d \leq 0$ and $0 < z/d \leq 1$ can be clearly seen in $|\Delta(z)|$ at any value of $1/k_F a_s$. On the other hand, especially in the BCS side ($1/k_F a_s = -1$), there is no large difference in $n(z)$ between the two regions [see the red line in figure 1(b)]. Going to the deep BCS regime, the period-2 nature is mainly possessed by the pairing field, which decreases to zero. This reflects the fact that our period-doubled states emerge due to the superfluidity. Furthermore, we have found that the period-doubled states become energetically more stable compared to the normal Bloch states in the BCS regime (see figure 2). This is in sharp contrast to the situation of the period-doubled states in BECs and in the BEC side of the BCS-BEC crossover.

In the deep BCS regime, the period-2 states are not only energetically more stable, but also they can be long-lived although dynamically unstable. We have studied the dynamical stability of the period-2 states by solving the time-dependent BdG equations [18]. The black solid line in figure 3 shows the growth rate γ of the fastest exponentially growing mode $|\eta(t)| = |\eta(0)|\, e^{\gamma t}$ of the deviation $|\Delta(x,t)| - |\Delta_0(x)|$ from the true stationary state $\Delta_0(x)$ for the period-2 states. We

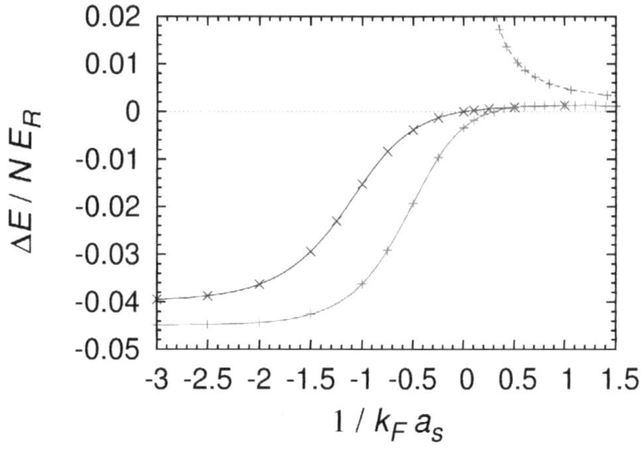

Figure 2. Difference ΔE of the energy per particle in units of E_R between the normal Bloch states and period-doubled states at $P = P_{\rm edge}/2$. Here we define $\Delta E \equiv E_2 - E_1$, where E_1 and E_2 represent the energy of the normal Bloch states and that of the period-doubled states, respectively. Parameters we have used are $s = 1, 2$ and $E_F/E_R = 0.25$. The red solid line with $+$ ($s = 1$) and blue solid line with $\times$ ($s = 2$) show the results obtained by solving the BdG equations and the green dashed line shows the results by the GP equation for corresponding parameters. Note that, in the BCS side, the energy of the period-doubled states at $P = P_{\rm edge}/2$ is lower than that of the normal Bloch states while the latter is lower than the former in the deep BEC side. This figure is taken from [18].

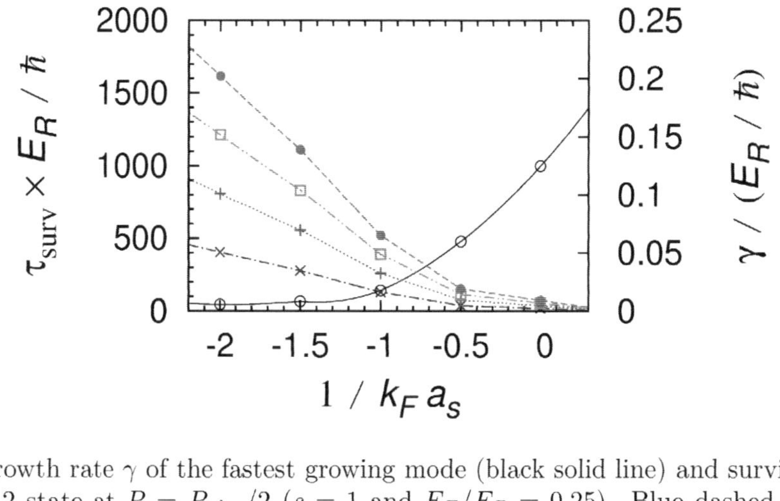

Figure 3. Growth rate γ of the fastest growing mode (black solid line) and survival time $\tau_{\rm surv}$ of the period-2 state at $P = P_{\rm edge}/2$ ($s = 1$ and $E_F/E_R = 0.25$). Blue dashed-dotted, green dotted, magenta dashed double-dotted, and red dashed lines show $\tau_{\rm surv}$ for relative amplitude $\tilde{\eta}(0)$ of the initial perturbation of 10%, 1%, 0.1%, and 0.01%, respectively. This figure is taken from [18].

see that γ is suppressed with decreasing $1/k_F a_s$, which makes the period-2 states long-lived in the BCS regime. The growth rate γ corresponds to the imaginary part of the complex eigenvalue for the fastest growing mode obtained by the linear stability analysis [23, 2], which is intrinsic property of the initial stationary state independent of the magnitude of the perturbation.

On the other hand, the actual survival time $\tau_{\rm surv}$, the timescale for which the initial state is destroyed by the large-amplitude oscillations, depends on the accuracy of their initial preparation. The survival time can be estimated by $\tilde{\eta}(0)e^{\gamma t} \sim 1$, where $\tilde{\eta}(0)$ is the relative amplitude of the initial perturbation with respect to $|\Delta_0|$ for the fastest growing mode. In figure 3, we show $\tau_{\rm surv}$ for various values of $\tilde{\eta}(0)$. This result suggests that if the initial stationary state is prepared within an accuracy of 10% or smaller, this state safely sustains for time scales of the order of $100\hbar/E_R$ or more in the BCS side, corresponding to $\tau_{\rm surv}$ of more than the order of a few milliseconds for typical experimental parameters [24]: For the recoil energy of bosonic molecules $E_{R,b} = 2\pi \times 7.3\text{kHz} \times \bar{}h$used in the experiment of [24], $1\hbar/E_R = 0.0109$ msec. In the deep BCS regime ($1/k_F a_s \ll -1$), $\tau_{\rm surv}$ increases further and may become larger than the time scale of the experiments, so that the period-doubled states can be regarded as long-lived states and, in addition, they have lower energy than the normal Bloch states in a finite range of quasimomenta. Therefore, by (quasi-)adiabatically increasing the quasimomentum P of the superflow starting from the ground state at $P = 0$, multiple-period states such as the period-doubled states could be realized experimentally in the deep BCS regime.

4. Summary and conclusion

Study of nonlinear phenomena in superfluid Fermi gases is a new research frontier. Indeed, the system of superfluid Fermi gas has richer physics compared to that of a Bose gas: Feshbach resonances enable us to explore the crossover between two limiting cases of the weakly coupled Bardeen-Cooper-Schrieffer (BCS) state and the Bose-Einstein condensate (BEC) state of tightly bound bosonic dimers. Furthermore, this system could provide implications for other interesting systems such as matter in neutron stars.

Emergence of stationary states whose period is a multiple of the lattice constant, i.e., multiple period states is a typical nonlinear phenomenon. In [18], we have studied multiple period states of superfluid Fermi gases in an optical lattice. By solving Bogoliubov-de Gennes equations, we have found that, in the BCS side of the crossover, the multiple period states can be energetically favorable compared to the normal Bloch states and their survival time against dynamical instability drastically increases, suggesting that these states can be accessible in current experiments [24].

Acknowledgments

This work was supported by the Zhejiang University 100 Plan, by the Junior 1000 Talents Plan of China, by IBS through Project Code (IBS-R024-D1), by the Max Planck Society, MEST of Korea, Gyeongsangbuk-Do, Pohang City, for the support of the JRG at APCTP, by NRF of Korea through the Basic Science Research Program (Grant No. 2012R1A1A2008028), by JSPS KAKENHI (Grant No. 24105006), by ERC through the QGBE grant, and by the Provincia Autonoma di Trento. Part of calculations were performed on RICC in RIKEN.

References

[1] Watanabe G, Venkatesh B P and Dasgupta R 2016 *Entropy* **18** 118
[2] Pethick C J and Smith H 2008 *Bose Einstein Condensation in Dilute Gases* 2nd edn (Cambridge: Cambridge University Press)
[3] Bloch I, Dalibard J and Zwerger W 2008 *Rev. Mod. Phys.* **80** 885
[4] Eagles D M 1969 *Phys. Rev.* **186** 456
[5] Leggett A J 1980 in *Modern Trends in the Theory of Condensed Matter* (Berlin: Springer-Verlag) pp 13–27
[6] Nozières P and Schmitt-Rink S 1985 *J. Low Temp. Phys.* **59** 195
[7] Sá de Melo C A R, Randeria M and Engelbrecht J R 1993 *Phys. Rev. Lett.* **71** 3202
[8] Giorgini S, Pitaevskii L P and Stringari S 2008 *Rev. Mod. Phys.* **80** 1215
[9] Machholm M, Nicolin A, Pethick C J and Smith H 2004 *Phys. Rev.* A **69** 043604
[10] Watanabe G, Yoon S, Dalfovo F 2011 *Phys. Rev. Lett.* **107** 270404

[11] Watanabe G and Yoon S 2013 *JKPS* **63** 839
[12] Yu D, Yi W and Zhang W 2015 *Phys. Rev.* A **92** 033623
[13] Gezerlis A and Carlson J 2008 *Phys. Rev.* C **77** 032801(R)
[14] Gezerlis A and Carlson J 2010 *Phys. Rev.* C **81** 025803
[15] Watanabe G, Dalfovo F, Pitaevskii L P and Stringari S 2011 *Phys. Rev.* A **83** 03321
[16] Ku M J H, Ji W, Mukherjee B, Guardado-Sanchez E, Cheuk L W, Yefsah T and Zwierlein M 2014 *Phys. Rev. Lett.* **113** 065301
[17] Ku M J H, Ji W, Mukherjee B, Yefsah T and Zwierlein M 2016 *Phys. Rev. Lett.* **116** 045304
[18] Yoon S, Dalfovo F, Nakatsukasa T and Watanabe G 2016 *New J. Phys.* **18** 023011
[19] de Gennes P G 1966 *Superconductivity of Metals and Alloys* (New York: Benjamin) ch 5 p 137
[20] Randeria M 1995 *Bose-Einstein Condensation* ed A Griffin *et al* (Cambridge: Cambridge University Press) ch 15 p 355
[21] Bruun G, Castin Y, Dum R and Burnett K 1999 *Eur. Phys. J.* D **7** 433
[22] Bulgac A and Yu Y 2002 *Phys. Rev. Lett.* **88** 042504
[23] Ring P and Schuck P 1980 *The Nuclear Many-Body Problem* (New York: Springer)
[24] Miller D E, Chin J K, Stan C A, Liu Y, Setiawan W, Sanner C and Ketterle W 2007 *Phys. Rev. Lett.* **99** 070402

Coherence Trade-off relations in Multipartite systems

Chandrashekar Radhakrishnan[1,2], Manikandan Parthasarathy[3], Segar Jambulingam[3] and Tim Byrnes[1,2,4,5].

[1] New York University, 1555 Century Avenue, Pudong, Shanghai 200122, China
[2] NYU-ECNU Institute of Physics at NYU Shanghai, 3663 Zhongshan Road North, Shanghai 200062, China
[3] Department of Physics, Ramakrishna Mission Vivekananda College, Mylapore, Chennai 600004, India
[4] Department of Physics, New York University, New York 10002, USA
[5] National Institute of Informatics, 2-1-2 Hitotsubashi, Chiyoda-ku, Tokyo 101-8430, Japan

E-mail: `cr2442@nyu.edu`

Abstract. Quantum coherence is investigated using a new measure with metric properties and entropic nature and decomposed into local and intrinsic contributions. The trade-off relation between these contributions as well as their distribution properties are studied for simple tripartite systems and the more complex spin chain model. We find that the coherence changes its nature with respect to the parameters of the quantum state under investigation.

1. Introduction

Quantifying coherence in the framework of quantum information theory was introduced in [1] and further developed in [2, 3]. The fundamental definitions of incoherent states, incoherent operations and maximally coherent states and the set of properties a functional should satisfy to be considered as a coherence measure were discussed in [1]. All these developments, were necessitated due to the importance of quantum coherence in a wide variety of fields [4, 5, 6, 7]. Coherence is a basic quantum property like entanglement, but it differs from entanglement on the basis of two features *viz*

(i) Coherence is a basis dependent quantity.
(ii) Coherence can arise due to intra-qubit nature as well as inter-qubit correlations.

In this work we delve on the second feature and formulate measuring schemes to estimate the two different forms of coherence originating from the inter-qubit and intra-qubit nature, as well as the total coherence of the system. Further we investigate how coherence is shared between the subsystems in multipartite systems. Distribution of coherence is discussed through the introduction of monogamy type of relations for coherence. Finally, these concepts are applied to study the Heisenberg XXZ-spin chain model.

Two different coherence measures namely the relative entropy of coherence and the ℓ_1 norm were used in [1]. While the relative entropy is an entropic measure, the ℓ_1-norm is geometric in nature with formal distance properties. A function d over a set X is called a distance if $\forall x, y \in X$ it satisfies the properties *(i)* $d(x,y) > 0 \; \forall \; x = y$ and $d(x,x) = 0$ (Positivity) *(ii)* $d(x,y) = d(y,x)$ (Symmetry). In addition if d satisfies $d(x,y) + d(y,z) \geq d(x,z)$ i.e., the triangle inequality, then d is a metric over the space X. We can immediately see that the relative entropy is not symmetric and hence is not a distance, but the ℓ_1 measure is a distance since it satisfies the axioms described above. A new coherence measure with both entropic and geometric nature is introduced using the quantum version of the Jensen-Shannon divergence (QJSD):

$$\mathcal{J}(\rho, \sigma) = \frac{1}{2}[S(\rho\|(\rho+\sigma)/2) + S(\sigma\|(\rho+\sigma)/2)]. \tag{1}$$

The QJSD is a distance measure and bounded $0 \leq \mathcal{J} \leq 1$, but it does not obey the triangle inequality. On the contrary, the square root of QJSD is a distance measure obeying the triangle inequality (metric) for all pure states [8, 9]. Though there is no proof that the mixed states obey triangle inequality, numerical studies up to five qubits [10] indicate its validity.

2. Local, Intrinsic and Total coherence: Definition and Properties

The basic properties a functional $\mathcal{C}$ should satisfy to be considered as a quantum coherence are as follows: *(i)* $\mathcal{C} \geq 0$ and $\mathcal{C} \equiv 0$ iff $\rho \in \mathcal{I}^{(b)}$, where $\mathcal{I}^{(b)}$; is the set of incoherent states. *(ii)* $\mathcal{C}(\rho)$ is invariant under unitary transformations, *(iii)* $\mathcal{C}(\rho)$ is monotonic under incoherent completely positive trace preserving (ICPTP) map, as well as under selective incoherent measurements on average. *(iv)* $\mathcal{C}(\rho)$ is convex i.e., does not increase under mixing of quantum states. In terms of a distance measure the coherence in an arbitrary state is the distance to the closest incoherent state

$$\mathcal{C}(\rho) \equiv \min_{\sigma\in\mathcal{I}^{(b)}} \mathcal{D}(\rho, \sigma), \tag{2}$$

In general the incoherent state $\sigma = \sum_k p_k|b_k\rangle\langle b_k|$ where $|b_k\rangle$ are the fixed basis and p_k are the probabilities. For a multipartite system, the incoherent state $\sigma = \sum_k p_k \tau_{k,1}^{(b)} \otimes \cdots \otimes \tau_{k,N}^{(b)}$,where $\tau_{k,n}^{(b)}$ is the incoherent state on the subsystem n i.e., $\tau_{k,n}^{(b)} = \sum_k p_{k,n}|b_{k,n}\rangle\langle b_{k,n}|$.

At this juncture we need to notice in a bipartite system, there are two types of coherent states *viz*, $(|0\rangle - |1\rangle)(|0\rangle - |1\rangle)$ and $|0\rangle|0\rangle - |1\rangle|1\rangle$. In the former type the coherence is on each qubit and in the latter one it is collective in nature. A schematic sketch describing the local coherence, intrinsic coherence and the total coherence are given in Fig. 1(a). The intrinsic coherence can be computed by relaxing the basis constraint. The definition of intrinsic coherence is

$$\mathcal{C}_I(\rho) \equiv \min_{\sigma_S\in\mathcal{I}_S} \mathcal{D}(\rho, \sigma_S), \tag{3}$$

where $\mathcal{I}_S$ is the set of incoherent state, and it is not in the same basis as ρ. Since the contribution is independent of the basis choice, it is intrinsic in nature and it is equal to the entanglement in the system. The coherent that is localized on the individual qubits - local coherence can be comptued using the relation

$$\mathcal{C}_L(\rho) \equiv \mathcal{D}(\sigma_S^{\min}, \rho^d), \tag{4}$$

where $\sigma_S^{\min}$ and ρ^d are the minimum solutions of (3) and (2) respectively. Based on the triangle inequality we can notice that

$$\mathcal{C} \leq \mathcal{C}_L + \mathcal{C}_I. \tag{5}$$

A very nice example corroborating the coherence decompositions is given by the ground state of the $N = 2$ Ising model described by the Hamiltonian

$$H = \lambda\sigma_1^x\sigma_2^x + J(\sigma_1^x + \sigma_2^x) + \epsilon\lambda(\sigma_1^z + \sigma_2^z). \tag{6}$$

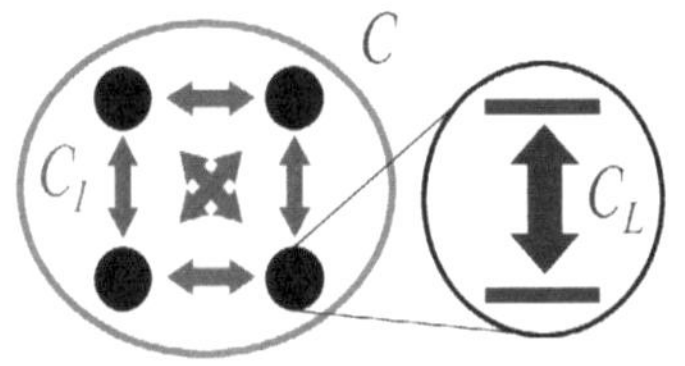

Figure 1. Quantum coherence in multipartite system. (a) The total coherence $\mathcal{C}$ with contributions from local coherence $\mathcal{C}_L$ and the Intrinsic coherence $\mathcal{C}_I$. (b) Definition of various coherences as the distanc between the states.

The parameters J and λ are the the coupling parameters and ϵ is a symmetry breaking term. Using the square root of QJSD defined via the expression

$$\mathcal{D}(\rho,\sigma) = \sqrt{\mathcal{J}(\rho,\sigma)} = \sqrt{S\left(\frac{\rho+\sigma}{2}\right) - \frac{S(\rho)}{2} - \frac{S(\sigma)}{2}},$$

we numerically estimate the values of C_L, C_I and C and the results are given in Fig 2 (a). We observe that the local coherence and the intrinsic coherence are complementary to each other. In the limit $J \ll \lambda$ the coherence is intrinsic in nature since the ground state approaches a Bell state in the $J = 0$ and $\epsilon \to 0$ limit. In the limit $J \ll \lambda$ the coherence is localized with each spin since for $\lambda = 0$ the ground state is $(|0\rangle - |1\rangle)(|0\rangle - |1\rangle)$. Also we notice that the total coherence is less than the sum of the local and intrinsic coherence as suggested in Eq. (5).

3. Shareability and Distribution of Quantum coherence:

The shareability and distribution of quantum coherence in multipartite system is explored in this section. The tripartite coherence in a system ρ_{123} may be decomposed in any one of the following forms

$$\mathcal{C}_{123} \leq \mathcal{C}_1 + \mathcal{C}_2 + \mathcal{C}_3 + \mathcal{C}_{1:2:3}, \qquad \mathcal{C}_{123} \leq \mathcal{C}_1 + \mathcal{C}_2 + \mathcal{C}_3 + \mathcal{C}_{2:3} + \mathcal{C}_{1:23}. \tag{7}$$

where C_n is the local coherence of the nth subsystem obtained from the reduced density matrix ρ_n and $C_{1:2:3}$ is the intrinsic coherence i.e., $C_I(\rho_{123})$. There are many such equivalent decompositions from which can conclude that

$$\mathcal{C}_{1:2:3} \simeq \mathcal{C}_{2:3} + \mathcal{C}_{1:23} \simeq \mathcal{C}_{1:2} + \mathcal{C}_{12:3} \simeq \mathcal{C}_{1:3} + \mathcal{C}_{13:2}. \tag{8}$$

Eq. (8) gives us a basic idea about how quantum coherence is shared between the qubits. To understand the distribution of coherence we naturally need to define the monogamy of coherence similar to the monogamy of entanglement [11, 12]. In a maximally coherent tripartite system ρ_{123}, the coherence between the system 1 and the bipartition 23 is related to the coherence between the substems 1 and 2 as well the coherence between 1 and 3 through the inequality

$$C_{1:23} \geq C_{1:2} + C_{1:3}. \tag{9}$$

If the inequality is obeyed the system is called monogamous and if not it is referred to as polygamous system. For a multipartite system the inequality is $C_{1:2...N} \geq \sum_{n=1}^{N} C_{1:n}$ and we define the measure

$$M = \sum_{n=2}^{N} C_{1:n} - C_{1:2...N} \tag{10}$$

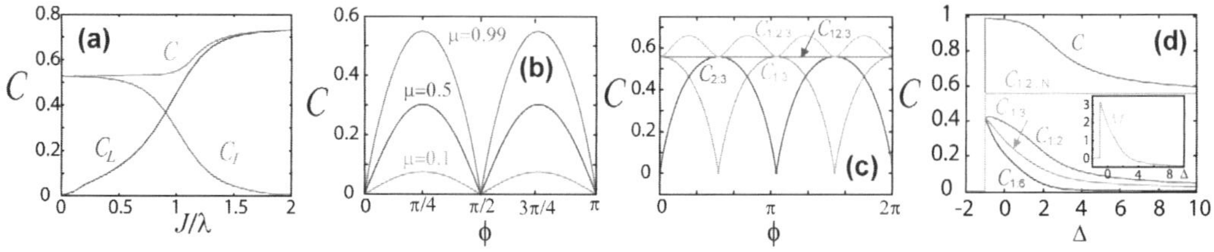

Figure 2. Coherence measured using QJSD for (a) $N = 2$ site Ising model with $\epsilon = 0.2$; (b) Werner GHZ state (c) W state with $\theta = \pi/4$; (d) $N = 10$ site XXZ spin chain model with $J = 1$. Inset: Monogamy of the XXZ spin chain.

which is monogamous for $M \leq 0$ and polygamous $M > 0$. Thus from the monogamy concept we get to know whether the coherence is distributed in a bipartite fashion or in a multipartite fashion.

4. Application to Multipartite systems

4.1. Tripartite systems

The tripartite states can be divided into the GHZ and the W class which are unrelated under local operations and classical communication [13]. *(i)* GHZ class: The coherence of the mixed GHZ states defined via the expression $\rho_{\text{GHZ}} = \frac{1-\mu}{8}\hat{\mathbb{1}} + \mu\ |\text{GHZ}\rangle\langle\text{GHZ}|$ where $|\text{GHZ}\rangle = \cos\phi|000\rangle + \sin\phi|111\rangle$ is plotted in Fig 2(b) for $\phi \in [0, 2\pi)$ and $0 \leq \mu \leq 1$. From the plot we observe that the local coherence C_n and the bipartitions $C_{m:n}$ is always zero. Hence we conclude that the intrinsic coherence in this system pertains to situations where all the three sites are involved. The total coherence is equal to $C = C_{1:2:3}$ and is also equal to the bipartition $C_{l:mn}$. *(ii)* W class: The generalized W state is defined as $|W\rangle = \sin\theta\cos\phi|100\rangle + \sin\theta\sin\phi|010\rangle + \cos\theta|001\rangle$ with $0 \leq \phi < 2\pi$ and $0 \leq \theta\pi$. In Fig 2(c) we plot the coherences from the various contributions corresponding to $\theta = \pi/4$. We notice that the coherences $C_{1:3}$ and $C_{2:3}$ show complementary behaviour as the system oscillates between the maximally entangled Bell state between sites 13 $[\phi = n\pi]$ and 23 $[\phi = (n + 1/2)\pi]$ and the third site is always decoupled. The intrinsic coherence $C_{12:3}$ is always a constant. Thus we notice that the W states show trade-off relations analogous to the Ising model. The GHZ states which are coherently connected in a tripartite manner have a monogamous behaviour whereas the W state which has a bipartite nature of coherence, exhibits polygamy over the entire range of the generalization parameters.

4.2. Spin Chain

The nature of quantum states in many-body systems can be understood from the study of quantum coherence in the systems. Towards this end we investigate the one dimensional Heisenberg XXZ model whose Hamiltonian is

$$H = J\sum_n(\sigma_n^x\sigma_{n+1}^x + \sigma_n^y\sigma_{n+1}^y + \Delta\sigma_n^z\sigma_{n+1}^z), \tag{11}$$

where J is the nearest neighbor spin coupling and Δ is the anisotropy parameter. When $J > 0$, the system exhibits a phase transition from the ferromagnetic axial regime to the antiferromagnetic regime at $\Delta = -1$. We estimate the various types of coherences as shown in Fig 2(d) using exact diagonalization techniques. The ferromagnetic ground state is unique and so in the ferromagnetic $(\Delta < -1)$ limit all the coherences vanish. In the limit $\delta \gg 1$ the ground state is a superposition of the Neel states $(|0101...01\rangle + |1010...10\rangle)/\sqrt{2}$ and coherence saturates to the Bell state value $C = C_{1:2...N} \sim 0.56$. The local coherence C_n is always zero due the spin flip symmetry and the intrinsic coherence between any two sites $C_{1:n}$ decreases with the increase in distance between them. But at $\Delta = -1$ they all converge to the same value since phase transition increases the overall effect of coherence in the system.

The Heisenberg spin chain exhibits monogamous behaviour for $\Delta > 2.9$ and polygamous behaviour $-1 < \Delta < 2.9$ as shown in Fig 2(d) inset. In the $\Delta \gg 1$ limit when the ground state is a Néel state the coherence is entirely due the distribution $C_{1:2...N}$. This is due to the fact that the Néel state is the same as the GHZ state. The dominant off-diagonal terms $\sigma_n^x\sigma_{n+1}^x + \sigma_n^y\sigma_{n+1}^y = 2(\sigma_n^+\sigma_{n+1}^- + \sigma_n^-\sigma_{n+1}^+)$ tend to create coherence on nearby sites when Δ is small and hence the system displays polygamous behaviour in this limit. By redistributing quantum coherence from the relatively local sites to the genuninely multipartite form the anisotropy parameter switches the spin chain between polygamous and monogamous nature.

5. Conclusions

Quantum coherence is quantified using a distance measure with metric properties which enables its decomposition into Local and Intrinsic contributions. The nature of coherence enables us to determine the type of state under consideration i.e., whether it is a genuinely multipartite state or more bipartite in nature. The coherence transitions from local to intrinsic nature in the Ising model. In the case of Heisenberg XXZ model, there is a cross over from the monogamous to the polygamous behaviour as the anisotropy parameter is varied. Application of these concepts to the fields of condensed matter physics and quantum metrology will further our understanding of quantum phase transition [14], frustration effects in magnetism [15] and may also lead to interferometric advantages [16, 17].

6. Acknowledgments

The work is supported by the Shanghai Research Challenge Fund, New York University Global seed Grants for Collaborative Research, National Natural Science Foundation of China Grant No. 61571301, and the Thousand Talents Program for Distinguished Young Scholars.

References

[1] Baumgratz T, Cramer M and Plenio M B 2014 *Phys. Rev. Lett* **113** 140401
[2] Girolami D 2014 *Phys. Rev. Lett* **113** 170401
[3] Streltsov A, Singh U, Dhar H S, Bera M N and Adesso G 2015 *Phys. Rev. Lett* **115** 020403
[4] Narasimhachar V S and Gour G 2015 *Nature Communications* **6** 7689
[5] Lostaglio M, Korzekwa K, Jennings D and Rudolph T 2015 *Phy. Rev.* X **5** 021001
[6] Engel G S 2007 *Nature* **446** 782
[7] Romero E et. al 2014 *Nature Physics* **10** 676
[8] Briet J and Harremoes P 2009 *Phys. Rev.* A **79** 052311
[9] Majtey A P, Lamberti P W and Prato D P 2005 *Phys. Rev.* A **72** 052310
[10] Lamberti P W, Majtey A P, Borras A, Casas M and Plastino A 2008 *Phys. Rev.* A **77** 052311
[11] Coffman V, Kundu J and Wootters W K 2000 *Phys. Rev.* A **61** 052306
[12] Osborne T J and Verstraete F 2006 *Phys. Rev. Lett.* **96** 220503
[13] Dür W, Vidal G and Cirac J I 2000 *Phys. Rev.* A **62** 062314
[14] Buluta I and Nori F 2009 *Science* **326** 108
[15] Ferraro A, García-Saez A and Acín A 2007 *Phys. Rev.* A **76** 052321
[16] Higgins B L, Barry D W, Bartlett S D, Wiseman H M and Pyrde G J 2007 *Nature* **450** 393
[17] Sahota J and Quesada N 2015 *Phys. Rev.* A **91** 013808

International Conference on the Frontiers in Atomic, Molecular, and Optical Physics (AMO2016) IOP Publishing
Journal of Physics: Conference Series **752** (2016) 012004 doi:10.1088/1742-6596/752/1/012004

STIRAP schemes for atomic Bose-Einstein condensates

Andreas M. D. Thomasen* and Tim Byrnes**

*,** New York University, 1555 Century Avenue, Pudong, Shanghai 200122, China
** National Institute of Informatics, 2-1-2 Hitotsubashi, Chiyoda-ku, Tokyo 101-8430, Japan
** NYU-ECNU Institute of Physics at NYU Shanghai, 3663 Zhongshan Road North, Shanghai 200062, China

E-mail: andreas.thomasen@nyu.edu

Abstract. In coherent control of Bose-Einstein condensates (BEC) the major limitation to successful high fidelity transfer using optical Raman transitions is spontaneous emission of radiation and the decoherence that this causes. We present a scheme based on stimulated Raman adiabatic passage (STIRAP) designed specifically for Rubidium 87. STIRAP is a method of population transfer that relies on the adiabatic theorem of quantum mechanics. Instead of transferring population between Hamiltonian ground states, one adiabatically evolves the ground states into some desired superposition. We show that one may thereby implement arbitrary rotations of spinor BECs with extremely high fidelities. Our simulations show that given a BEC with $N = 10^4$ atoms, we may do an arbitrary unitary rotation with an infidelity of about 10^{-7} that takes approximately 100 ns.

1. Introduction

Two component Bose-Einstein condensates may exist in coherent states of matter that are accurately described by

$$|\alpha, \beta\rangle\rangle \equiv \frac{1}{\sqrt{N!}}\left(\alpha \hat{a}^\dagger + \beta \hat{b}^\dagger\right)^N |0\rangle . \tag{1}$$

Here $a^\dagger$ and $b^\dagger$ create an atom in the BEC in some internal orthogonal states that are labeled a and b respectively. We propose that this type of state may be accurately controlled by a stimulated Raman adiabatic passage (STIRAP). Furthermore, we do so in the presence of bosonically enhanced spontaneous emission. Similar schemes have been proposed before such as [1, 2], and we base our ideas partly on them.

The main advantage of using STIRAP is that one avoids populating excited internal states of the BEC constituent atoms and thus limits decoherence caused by spontaneous emission [3, 4].

1.1. STIRAP

STIRAP is a thoroughly studied and well-understood method of population transfer in atoms and molecules [5, 6, 7]. A time-dependent Hamiltonian has some set of instantaneous eigenstates $|d_1(t)\rangle$, $|d_2(t)\rangle$, . . . , $|d_n(t)\rangle$. A general quantum state may then be written as

$$|\psi(t)\rangle = \sum_j c_j(t) |d_j(t)\rangle . \tag{2}$$

In the adiabatic approximation each of the $c_j(t) \approx c_j(0)$.

In STIRAP, one typically defines *dark states* as the eigenstates that contain only stable electronic states, i.e. states that do not spontaneously emit radiation. The ones that do are termed *bright states*. Consider a superposition of two dark states as the initial state of some atomic system, then an elementary STIRAP procedure is one in which the two dark states replace each other. I.e.

$$\begin{aligned} |\psi(t)\rangle &= c_1(0)\,|d_1(t)\rangle + c_2(0)\,|d_2(t)\rangle \\ &\to c_1(0)\,|d_2(t)\rangle + c_2(0)\,|d_1(t)\rangle\,. \end{aligned} \tag{3}$$

This is typically accomplished by a so-called counter-intuitive sequence of laser pulses, where a Stokes pulse precedes a pump pulse. Thus no Rabi transition occurs and only dark states are ever populated in the adiabatic limit.

As an example of how this occurs, consider the Hamiltonian

$$\hat{H}/\hbar = [\Omega_a(t)\hat{e}^\dagger\hat{a} + \Omega_b(t)\hat{e}^\dagger\hat{b} + \text{H.c.}] - \Delta/2\hat{a}^\dagger\hat{a} - \Delta/2\hat{b}^\dagger\hat{b} + \Delta/2\hat{e}^\dagger\hat{e}. \tag{4}$$

Here the dark states may be created by the operator

$$\hat{d}_0^\dagger = \frac{\Omega_b(t)\hat{a}^\dagger - \Omega_a(t)\hat{b}^\dagger}{\sqrt{|\Omega_a(t)|^2 + |\Omega_b(t)|^2}}. \tag{5}$$

Assuming real valued non-negative laser amplitudes, we may define the mixing angle $\theta(t)$ as

$$\frac{\Omega_a}{\cos\theta(t)} = \frac{\Omega_b}{\sin\theta(t)}, \quad 0 \le \theta(t) \le \pi/2. \tag{6}$$

Then

$$\hat{d}_0^\dagger = \cos\theta(t)\hat{a}^\dagger - \sin\theta(t)\hat{b}^\dagger. \tag{7}$$

Now the mixing angle defines some adiabatic path taken by the eigenstate, which may transfer population completely from a to b or b to a, as well as ending the transfer in some superposition of the two (fractional STIRAP).

1.2. Internal state manifolds of ^{87}Rb

We consider specifically ^{87}Rb BECs. Here the $5^2S_{1/2}$ and $5^2P_{1/2}$ manifolds of hyperfine states are the most relevant optically addressable states. In the rotating frame we may write a Hamiltonian as

$$\begin{aligned} \hat{H}/\hbar = &\left[\Omega_a(t)\left(\hat{e}_1^\dagger + \hat{e}_2^\dagger\right)\hat{a} + \Omega_b(t)\left(\hat{e}_1^\dagger - \hat{e}_2^\dagger\right)\hat{b} + \Omega_{c_1}e_1^\dagger\hat{c}_1 + \Omega_{c_2}e_2^\dagger\hat{c}_2 + \text{H.c.}\right] \\ &+ \Delta_1\hat{e}_1^\dagger\hat{e}_1 + \Delta_2\hat{e}_2^\dagger\hat{e}_2. \end{aligned} \tag{8}$$

Here $\hat{a}^\dagger$, $\hat{b}^\dagger$, $\hat{c}_1^\dagger$ and $\hat{c}_2^\dagger$ create an atom in the hyperfine ground states $|F=1, m_F=-1\rangle$, $|F=2, m_F=1\rangle$, $|F=2, m_F=0\rangle$ and $|F=1, m_F=0\rangle$ respectively of the $5^2S_{1/2}$ manifold. The excited states $|F'=1, m_F'=0\rangle$ and $|F'=2, m_F'=0\rangle$ of the $5^2P_{1/2}$ manifold are created by $\hat{e}_1^\dagger$ and $\hat{e}_2^\dagger$ respectively.

In writing up this Hamiltonian we have used a rotating frame, where some of the off-diagonal elements have terms that rotate with a frequency given by the hyperfine splitting of the ground state. This is approximately 6.8 GHz, which is much faster than the timescales we consider. These terms therefore average to zero approximately.

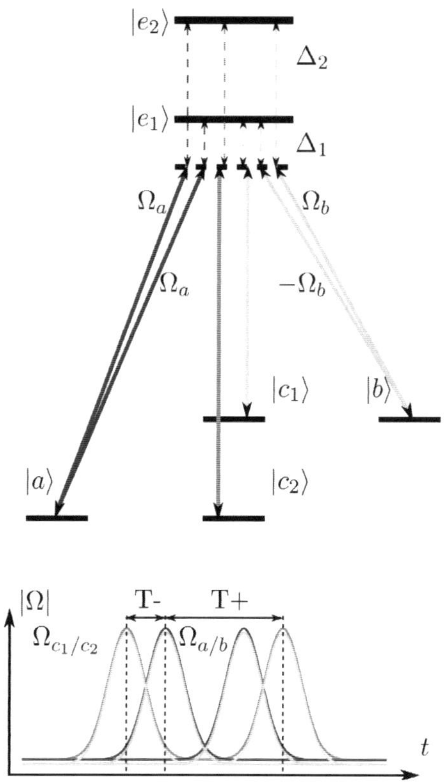

Figure 1. Here is a level scheme showing the internal electronic levels of Rubidium 87 (a). Also indicated are the STIRAP pulses and their relative timing in our scheme (b). Due to Klebsch Gordan coefficients specific to the D1 transition of ^{87}Rb the laser amplitudes Ω_a and Ω_b cause transitions to e_1 and e_2 to occur with different phases.

Equation (8) has the following dark states creation operators.

$$\hat{d}_1^{(1)\dagger} = \frac{\Omega_{c_1}\Omega_{c_2}\hat{a}^\dagger - \Omega_a\Omega_{c_2}\hat{c}_1^\dagger - \Omega_a\Omega_{c_1}\hat{c}_2^\dagger}{\sqrt{|\Omega_{c_1}|^2|\Omega_{c_2}|^2 + |\Omega_a|^2\left(|\Omega_{c_1}|^2 + |\Omega_{c_2}|^2\right)}} \tag{9}$$

$$\hat{d}_2^{(1)\dagger} = \frac{\Omega_{c_1}\Omega_{c_2}\hat{b}^\dagger - \Omega_b\Omega_{c_2}\hat{c}_1^\dagger + \Omega_b\Omega_{c_1}\hat{c}_2^\dagger}{\sqrt{|\Omega_{c_1}|^2|\Omega_{c_2}|^2 + |\Omega_b|^2\left(|\Omega_{c_1}|^2 + |\Omega_{c_2}|^2\right)}}. \tag{10}$$

We also define an equivalent dark state basis for later convenience. It is defined by the creation operators

$$\hat{d}_1^{(2)\dagger} = \frac{-\Omega_{c_1}\Omega_b\hat{a}^\dagger - \Omega_{c_1}\Omega_a\hat{b}^\dagger + \Omega_a\Omega_b\hat{c}_1^\dagger}{\sqrt{|\Omega_a|^2|\Omega_b|^2 + |\Omega_{c_1}|^2\left(|\Omega_a|^2 + |\Omega_b|^2\right)}} \tag{11}$$

$$\hat{d}_2^{(2)\dagger} = \frac{-\Omega_{c_2}\Omega_b\hat{a}^\dagger + \Omega_{c_2}\Omega_a\hat{b}^\dagger + \Omega_a\Omega_b\hat{c}_2^\dagger}{\sqrt{|\Omega_a|^2|\Omega_b|^2 + |\Omega_{c_2}|^2\left(|\Omega_a|^2 + |\Omega_b|^2\right)}}. \tag{12}$$

(13)

2. Arbitrary rotation of coherent spinors

We now may introduce a scheme that uses STIRAP for arbitrary rotation of BECs. This scheme will consist of two STIRAPs, where the first one takes the condensate into a coherent spinor composed of the intermediate $\hat{c}_1^\dagger$ and $\hat{c}_2^\dagger$ states. After this the second STIRAP will take the

Journal of Physics: Conference Series **752** (2016) 012004 doi:10.1088/1742-6596/752/1/012004

condensate back into the original subspace. This scheme is heavily inspired by Ref. [2]. In summary

$$
\begin{aligned}
|\alpha_0, \beta_0\rangle\rangle &= \frac{1}{\sqrt{N!}}\left(\alpha_0 \hat{a}^\dagger + \beta_0 \hat{b}^\dagger\right)^N |0\rangle && (14)\\
&\to \frac{1}{\sqrt{N!}}\left(\alpha_1 \hat{c}_1^\dagger + \beta_1 \hat{c}_2^\dagger\right)^N |0\rangle && (15)\\
&\to \frac{1}{\sqrt{N!}}\left(\alpha_2 \hat{a}^\dagger + \beta_2 \hat{b}^\dagger\right)^N |0\rangle . && (16)
\end{aligned}
$$

We may define the total transformation as an SU(2) rotation. Then

$$
\begin{bmatrix} \alpha_2 \\ \beta_2 \end{bmatrix} = U \begin{bmatrix} \alpha_0 \\ \beta_0 . \end{bmatrix} \tag{17}
$$

The main result of this work is that we have found a way to perform arbitrary rotations. It is well know that a general SU(2) rotation may be written as

$$
U = e^{i\alpha} \begin{bmatrix} e^{i(-\beta/2-\delta/2)} \cos\frac{\gamma}{2} & -e^{i(-\beta/2+\delta/2)} \sin\frac{\gamma}{2} \\ e^{i(\beta/2-\delta/2)} \sin\frac{\gamma}{2} & e^{i(\beta/2+\delta/2)} \cos\frac{\gamma}{2} \end{bmatrix}. \tag{18}
$$

We now prove that this is possible using a pair of STIRAPs on the internal state manifold of ^{87}Rb. Firstly, we define a unimodal function $f(t)$ whose maximum is $f(0) = 1$. In STIRAP this function is usually taken to be Gaussian. We now define T_- as the time difference between a Stokes and pump pulse and T_+ the time difference between the two STIRAP procedures. We then define explicitly the pulses

$$
\begin{aligned}
\Omega_a^{(1)} &= e^{i\theta_a^{(1)}} f(t - T_-/2 + T_+/2), & \Omega_a^{(2)} &= e^{i\theta_a^{(2)}} f(t + T_-/2 - T_+/2)\\
\Omega_b^{(1)} &= e^{i\theta_b^{(1)}} f(t - T_-/2 + T_+/2), & \Omega_b^{(2)} &= e^{i\theta_b^{(2)}} f(t + T_-/2 - T_+/2)\\
\Omega_{c_1}^{(1)} &= e^{i\theta_{c_1}^{(1)}} f(t + T_-/2 + T_+/2), & \Omega_{c_1}^{(2)} &= e^{i\theta_{c_1}^{(2)}} f(t - T_-/2 - T_+/2)\\
\Omega_{c_2}^{(1)} &= e^{i\theta_{c_2}^{(1)}} f(t + T_-/2 + T_+/2), & \Omega_{c_2}^{(2)} &= e^{i\theta_{c_2}^{(2)}} f(t - T_-/2 - T_+/2).
\end{aligned} \tag{19}
$$

Here the $\Omega_j^{(1,2)}$ are the pulses belonging to the first and second STIRAP respectively.

The dark states have the following limits in the first STIRAP.

$$
\hat{d}_1^{(1)\dagger} \to \begin{cases} e^{i(\theta_{c_1}^{(1)}+\theta_{c_2}^{(1)})}\hat{a}^\dagger & \text{for } t \to -\infty \\ \frac{-e^{i(\theta_a^{(1)}+\theta_{c_2}^{(1)})}\hat{c}_1^\dagger - e^{i(\theta_a^{(1)}+\theta_{c_1}^{(1)})}\hat{c}_2^\dagger}{\sqrt{2}} & \text{for } t \to \infty \end{cases} \tag{20}
$$

$$
\hat{d}_2^{(1)\dagger} \to \begin{cases} e^{i(\theta_{c_1}^{(1)}+\theta_{c_2}^{(1)})}\hat{b}^\dagger & \text{for } t \to -\infty \\ \frac{-e^{i(\theta_b^{(1)}+\theta_{c_2}^{(1)})}\hat{c}_1^\dagger + e^{i(\theta_b^{(1)}+\theta_{c_1}^{(1)})}\hat{c}_2^\dagger}{\sqrt{2}} & \text{for } t \to \infty. \end{cases} \tag{21}
$$

In the second STIRAP we have

$$
\hat{d}_1^{(2)\dagger} \to \begin{cases} e^{i(\theta_a^{(2)}+\theta_b^{(2)})}\hat{c}_1^\dagger & \text{for } t \to -\infty \\ \frac{-e^{i(\theta_{c_1}^{(2)}+\theta_b^{(2)})}\hat{a}^\dagger - e^{i(\theta_{c_1}^{(2)}+\theta_a^{(2)})}\hat{b}^\dagger}{\sqrt{2}} & \text{for } t \to \infty \end{cases} \tag{22}
$$

$$
\hat{d}_2^{(2)\dagger} \to \begin{cases} e^{i(\theta_a^{(2)}+\theta_b^{(2)})}\hat{c}_2^\dagger & \text{for } t \to -\infty \\ \frac{-e^{i(\theta_{c_2}^{(2)}+\theta_b^{(2)})}\hat{a}^\dagger + e^{i(\theta_{c_2}^{(2)}+\theta_a^{(2)})}\hat{b}^\dagger}{\sqrt{2}} & \text{for } t \to \infty. \end{cases} \tag{23}
$$

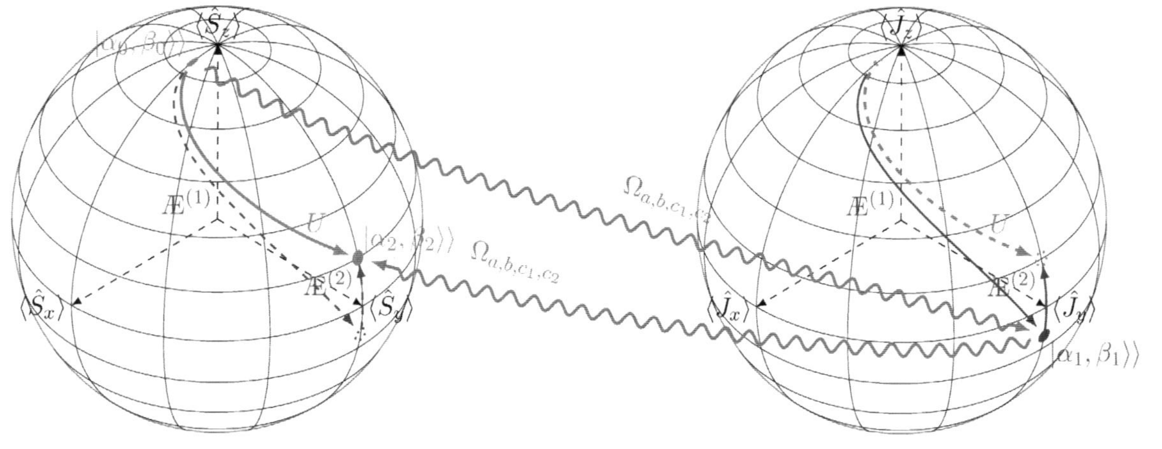

Figure 2. (Color online). This figure shows an interpretation of the overall scheme as a pair of complete transitions back and forth between two Bloch spheres. The rotation U is the net effect of these two transitions, each of which contribute to U with the partial rotations $Æ^{(1,2)}$. The wavy magenta arrows represent the action of the actual lasers in the STIRAP scheme, which is to make a transition happen from one Bloch sphere to another.

We now examine the first STIRAP given some initial state $|\alpha_0, \beta_0\rangle\rangle$.

$$|\alpha_0, \beta_0\rangle\rangle = \frac{1}{\sqrt{N!}}\left(\alpha_0 e^{-i(\theta_{c_1}^{(1)}+\theta_{c_2}^{(1)})}\hat{d}_1^{(1)\dagger} + \beta_0 e^{-i(\theta_{c_1}^{(1)}+\theta_{c_2}^{(1)})}\hat{d}_2^{(1)\dagger}\right)^N |0\rangle \tag{24}$$

$$\to \frac{1}{\sqrt{N!}}\left(\alpha_0 e^{-i(\theta_{c_1}^{(1)}+\theta_{c_2}^{(1)})}\frac{-e^{i(\theta_a^{(1)}+\theta_{c_2}^{(1)})}\hat{c}_1^\dagger - e^{i(\theta_a^{(1)}+\theta_{c_1}^{(1)})}\hat{c}_2^\dagger}{\sqrt{2}}\right.$$
$$\left. + \beta_0 e^{-i(\theta_{c_1}^{(1)}+\theta_{c_2}^{(1)})}\frac{-e^{i(\theta_b^{(1)}+\theta_{c_2}^{(1)})}\hat{c}_1^\dagger + e^{i(\theta_b^{(1)}+\theta_{c_1}^{(1)})}\hat{c}_2^\dagger}{\sqrt{2}}\right)^N |0\rangle \tag{25}$$

$$= |\alpha_1, \beta_1\rangle\rangle\,. \tag{26}$$

Here the coherent spinor is with respect to the intermediate states c_1 and c_2. Rearranging the terms of equation (26) according to equation (15), we obtain a matrix tranformation between $[\alpha_0, \beta_0]^T$ and $[\alpha_1, \beta_1]^T$ given by

$$Æ^{(1)} = \frac{1}{\sqrt{2}}\begin{bmatrix} -e^{i(\theta_a^{(1)}-\theta_{c_1}^{(1)})} & -e^{i(\theta_b^{(1)}-\theta_{c_1}^{(1)})} \\ -e^{i(\theta_a^{(1)}-\theta_{c_2}^{(1)})} & e^{i(\theta_b^{(1)}-\theta_{c_2}^{(1)})} \end{bmatrix}. \tag{27}$$

Now performing the second STIRAP procedure, the intermediate states move back into the original storage basis again.

$$|\alpha_1,\beta_1\rangle\rangle = \frac{1}{\sqrt{N!}}\left(\alpha_1 e^{-i(\theta_a^{(2)}+\theta_b^{(2)})}\hat{d}_1^{(2)\dagger} + \beta_1 e^{-i(\theta_a^{(2)}+\theta_b^{(2)})}\hat{d}_2^{(2)\dagger}\right)^N |0\rangle \tag{28}$$

$$\rightarrow \frac{1}{\sqrt{N!}}\left(\alpha_1 e^{-i(\theta_a^{(2)}+\theta_b^{(2)})}\frac{-e^{i(\theta_{c1}^{(2)}+\theta_b^{(2)})}\hat{a}^\dagger - e^{i(\theta_{c1}^{(2)}+\theta_a^{(2)})}\hat{b}^\dagger}{\sqrt{2}} + \beta_1 e^{-i(\theta_a^{(2)}+\theta_b^{(2)})}\frac{-e^{i(\theta_{c2}^{(2)}+\theta_b^{(2)})}\hat{a}^\dagger + e^{i(\theta_{c2}^{(2)}+\theta_a^{(2)})}\hat{b}^\dagger}{\sqrt{2}}\right)^N |0\rangle \tag{29}$$

$$= |\alpha_2,\beta_2\rangle\rangle\,. \tag{30}$$

Here the matrix that transforms $[\alpha_1,\beta_1]^T$ into $[\alpha_2,\beta_2]^T$ is given by

$$Æ^{(2)} = \frac{1}{\sqrt{2}}\begin{bmatrix} -e^{i(\theta_{c1}^{(2)}-\theta_a^{(2)})} & -e^{i(\theta_{c2}^{(2)}-\theta_a^{(2)})} \\ -e^{i(\theta_{c1}^{(2)}-\theta_b^{(2)})} & e^{i(\theta_{c2}^{(2)}-\theta_b^{(2)})} \end{bmatrix}. \tag{31}$$

We may now construct the total operation on the BEC as

$$U = Æ^{(2)}Æ^{(1)}. \tag{32}$$

By choosing laser phase angles $\theta_{c1}^{(1)} = \theta_{c2}^{(1)} = -\alpha/2$, $\theta_{c1}^{(2)} = \alpha/2+\gamma/2$ and $\theta_{c2}^{(2)} = \alpha/2-\gamma/2$, $\theta_a^{(1)} = -\delta/2$, $\theta_b^{(1)} = \delta/2+\pi/2$, $\theta_a^{(2)} = -\beta/2$, $\theta_b^{(2)} = \beta/2+\pi/2$, we obtain the arbitrary SU(2) rotation of equation (18).

Spin coherent states may be given a Bloch sphere representation by using the Schwinger Boson operators. They are

$$\begin{aligned} \hat{S}_x &= \hat{a}^\dagger\hat{b} + \hat{b}^\dagger\hat{a}, & \langle\hat{S}_x\rangle &= N(\alpha^*\beta + \beta^*\alpha) \\ \hat{S}_y &= -i\hat{a}^\dagger\hat{b} + i\hat{b}^\dagger\hat{a}, & \langle\hat{S}_y\rangle &= N(-i\alpha^*\beta + i\beta^*\alpha) \\ \hat{S}_z &= \hat{a}^\dagger\hat{a} - \hat{b}^\dagger\hat{b}, & \langle\hat{S}_z\rangle &= N\big(|\alpha|^2 - |\beta|^2\big). \end{aligned} \tag{33}$$

We may similarly define $\hat{J}_x = \hat{c}_1^\dagger\hat{c}_2 + \hat{c}_2^\dagger\hat{c}_1$, etc. Then an interpretation of the two STIRAP procedures as jumps from one Bloch sphere to another may be visualized as in figure 2.

3. Simulations

We have carried out a simulation in which

$$f(t) = e^{-t^2/T^2}. \tag{34}$$

This has been done for various values of N, Ω_0, t_{end}, where $t_{\text{end}} = 8T_- = 8T_+/3$. In each case, for a given set of those parameters, we have done the simulation in two steps. The first is an optimization procedure based on golden section search, in which an optimum value of the pulse width T is determined. In the second step, the simulation is carried out 240 times with this optimum T. Each time the initial state of the condensate is randomized in terms of α_0 and β_0. The angles of rotation in equation (18) are randomized as well. Each time a simulation finishes, the resulting final state ρ_{final} is compared to the target state ρ_{target} that one would have gotten simply by applying the unitary directly. The trace distance between the two is recorded and after all 240 simulations have been carried out, their average is saved for that combination of N, Ω_0, t_{end}. We note that in our simulation the master equation is procedurally evaluated from $t = 0$ to $t = t_{\text{end}}$. We also note that both T_- and T_+ are fixed with respect to t_{end}.

3.1. Master equation

We use the master equation

$$\frac{d\rho}{dt} = \frac{i}{\hbar}[\rho, \hat{H}] - \frac{\Gamma}{2}\left([\hat{e}_1^\dagger \hat{c}_1, \hat{c}_1^\dagger \hat{e}_1 \rho] + [\rho \hat{e}_1^\dagger \hat{c}_1, \hat{c}_1^\dagger \hat{e}_1]\right) - \frac{\Gamma}{2}\left([\hat{e}_2^\dagger \hat{c}_2, \hat{c}_2^\dagger \hat{e}_2 \rho] + [\rho \hat{e}_2^\dagger \hat{c}_2, \hat{c}_2^\dagger \hat{e}_2]\right). \tag{35}$$

Here Γ is the rate of spontaneous decay of the $5^2P_{1/2}$ excited state. We do not directly time-evolve the entire density matrix, as the size of the Fock-state space grows very fast with the number of particles, especially since we consider six interacting hyperfine states. Generally the number of Fock states equals the number of terms in an N'th order hexanomial expansion, which for all practical purposes is impossible to simulate in a reasonable amount of time with $N > 1000$.

We instead evaluate the expectation values of bosonic compound operators using the above master equation, i.e. $\langle \hat{a}^\dagger \hat{a} \rangle$, $\langle \hat{a}^\dagger \hat{b} \rangle$, ... , $\langle \hat{e}_2^\dagger \hat{e}_2 \rangle$. These form a self-consistent set of 16 coupled differential equations regardless of N. We have used a mean-field approximation, where terms such as $\langle \hat{a}^\dagger \hat{a} \hat{b}^\dagger \hat{b} \rangle \approx \langle \hat{a}^\dagger \hat{a} \rangle \langle \hat{b}^\dagger \hat{b} \rangle$. This approximation works well for spin coherent states, as the variance of spin operators diminishes with $1/N$ [8].

After deriving all relevant differential equations, we numerically solve it by using an ordinary differential equation (ODE) solver.

3.2. Evaluating fidelity

We use the trace distance as a measure of infidelity. We define the trace distance between two quantum states ρ and σ as

$$D(\rho, \sigma) = \left[\left(\langle \hat{S}_x \rangle_\rho - \langle \hat{S}_x \rangle_\sigma\right)^2 + \left(\langle \hat{S}_y \rangle_\rho - \langle \hat{S}_y \rangle_\sigma\right)^2 + \left(\langle \hat{S}_z \rangle_\rho - \langle \hat{S}_z \rangle_\sigma\right)^2\right]/(4N^2). \tag{36}$$

Then $F(\rho_{\text{target}}, \rho_{\text{final}}) \equiv 1 - D(\rho_{\text{target}}, \rho_{\text{final}})$. Our justification for this choice is that spin coherent states have a vanishing overlap which depends exponentially on N. This is given by

$$\begin{aligned} \left| \left\langle \left\langle \cos\frac{\theta'}{2}, e^{i\phi} \sin\frac{\theta'}{2} \,\middle|\, \cos\frac{\theta}{2}, e^{i\phi} \sin\frac{\theta}{2} \right\rangle \right\rangle \right|^2 &= \cos^{2N}\left(\frac{\theta' - \theta}{2}\right) \\ &\approx e^{-\frac{N(\theta'-\theta)^2}{4}}. \end{aligned} \tag{37}$$

Although the two spin coherent states may be very close together on the Bloch sphere, the overlap still vanishes. If a conventional approach was used where the fidelity is defined in terms of density matrix products, such expressions will inevitably appear.

3.3. Example Simulation

Figure 3 shows the time-evolution of two different spin coherent states. In the example shown, the detuning has been set to $\Delta_+ = 0$. The two graphs on the left show the time-evolution of a BEC with $N = 10^4$, while the right hand graphs have $N = 10^5$. We see that the BEC arrives at its target state in graph (a), whereas graph (b) shows some infidelity. This is because when the average detuning $\Delta_+ \equiv (\Delta_2 - \Delta_1)/2 = 0$. When lasers are near resonance, transitions to excited states are much more likely. Consequently we see a dependence on N emerge as superradiance becomes the main contributor to decoherence. This trend is mirrored in graphs (c) and (d). After the full procedure, no population is left in the intermediate states in (c), whereas in (d) some finite amount is left over. This shows directly that significant spontaneous emission has taken place.

This example shows that the counterintuitive pulse ordering of STIRAP effectively prevents excited states from being populated and thus overcomes enhanced spontaneous emission.

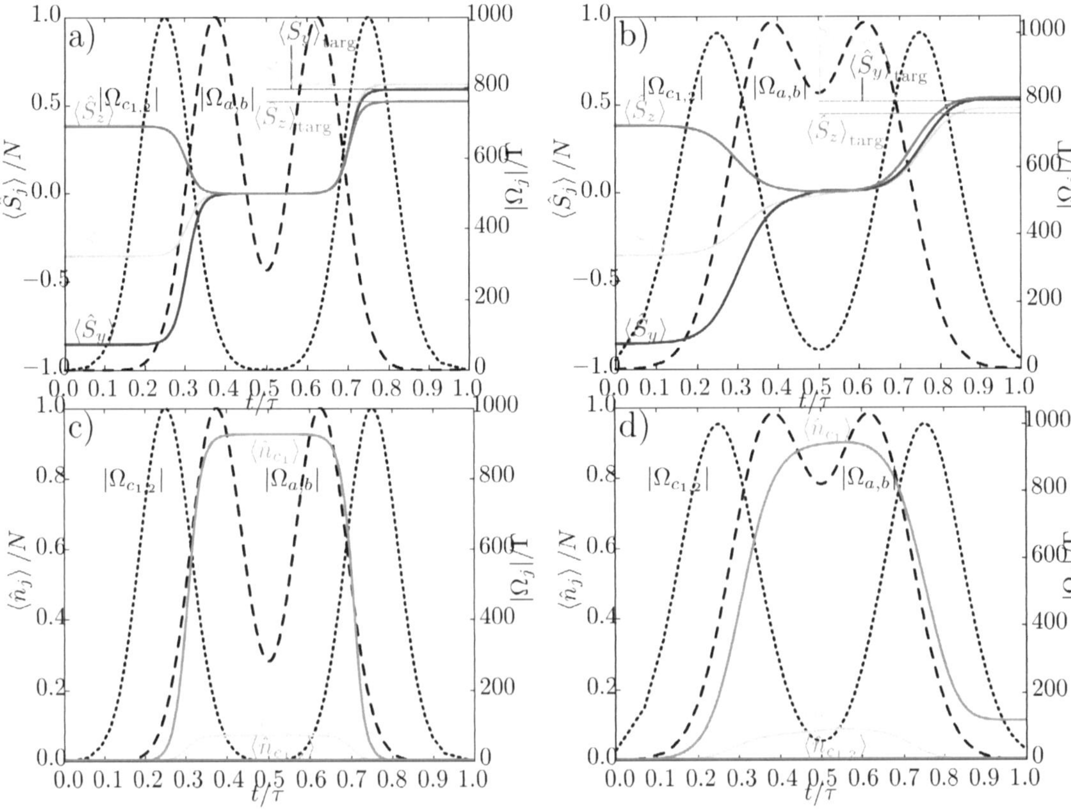

Figure 3. (Color online). Typical six-level STIRAP time evolution performing a coherent rotation of a BEC. The simulations have been carried out using data recorded for the D1 transition in [9], and the basic time unit is the decay time of the ^{87}Rb D1 transition, i.e. $\tau = 27.70(4)$ ns. The simulation ends after one τ. All plots are for the case where the initial spin coherent state is $|\cos 3\pi/8, \sin 3\pi/8e^{i5\pi/8}\rangle\rangle$ and the target rotation is the operation $U = \mathcal{R}_z(\pi/4)\mathcal{R}_y(3\pi/4)\mathcal{R}_z(\pi/4)$. The target state after the rotation is indicated with the horizontal lines marked as $\langle\hat{S}_{x,y,z}\rangle_{\rm targ}$. The pulses have a Gaussian shape, with $T_- = t_{\rm end}/8$, $T_+ = 3t_{\rm end}/8$. After pulse width optimization we obtain for (a) (c) $N = 10^4$ a width of $T = 0.1324t_{\rm end}$ (FWHM $= 0.1835t_{\rm end}$) and for (b) (d) $N = 10^5$ we obtain $T = 0.0877t_{\rm end}$ (FWHM $= 0.1216t_{\rm end}$). Subfigures (a) (b) show the the evolution of the expectation values of the Schwinger boson operators, $\langle\hat{S}_{x,y,z}\rangle$. Subfigures (c) (d) depict the occupation of intermediate and excited states.

However, in general there are more excited states in the $5^2P_{1/2}$ manifold of states than we've included in the Hamiltonian of equation (8). In a situation with low detuning, there will generally therefore be transitions to unwanted states. Therefore the above results should only be thought of as illustrative of the STIRAP procedure and the dependence of the fidelity on N due to enhanced spontaneous emission. In the following simulations we shall include higher values of Δ_+ to avoid these transitions entirely.

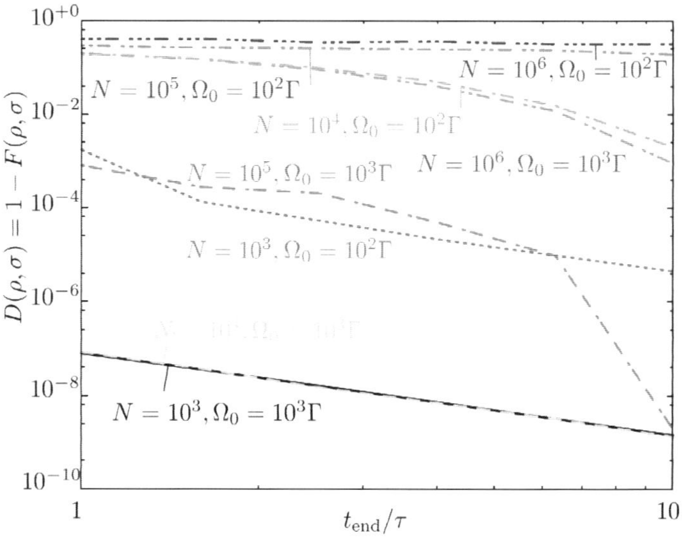

Figure 4. (Color online). Here the trace distance D is plotted as a function of t_{end}. This is done for eight data-series with different combinations of the parameters $N = 10^3, 10^4, 10^5, 10^6$ and $\Omega_0 = 10^2\Gamma, 10^3\Gamma$. For each combination of these values a datapoint has been computed at each t_{end} by averaging over the results of 240 randomized rotations and initial states.

3.4. Main results

Figure 3.4 shows our main results for a range of different values of the parameters N, Ω_0 and t_{end}. For any given set of parameters the time-evolution is carried out 240 times, each time as in figure 3, but with randomized initial states and unitary operations. The set of coupled expectation values of section 3.1 is procedurally evaluated from $t = 0$ to $t = t_{\text{end}}$ using an ODE solver. After the final time-step of the evolution has been carried out using this method, we characterize the state by evaluating the expectation values of equation (33). The trace distance is then taken to the target state, and all of the 240 values thus obtained are averaged over and saved for that particular configuration of N, Ω_0 and t_{end}. This process is for each set of parameters preceded by an optimization step in which the optimum width of the pulses T (see equation (34)) is obtained by a golden section search procedure. We note also that in each case, the detuning has been set to $\Delta_+ = 0$. This is to support our claim that STIRAP really does overcome the problem of spontaneous emission even very close to resonance.

On figure 3.4, which plots the infidelity vs t_{end}, several data-series have been plotted. There are eight data-series in total with $N = 10^3$, 10^4, 10^5, 10^6 and $\Omega_0 = 10^2\Gamma$, $10^3\Gamma$. Each data-series shows, as an overall trend, that allowing a longer t_{end} increases the fidelity. There is also an overall decrease in fidelity with increasing numbers of particles. Thus we find in general that there is a trade-off between N and t_{end} for some minimum required F. Most of the data-series with $\Omega_0 = 10^2\Gamma$ have infidelities close enough to 1 that most practical applications are difficult to justify. When $N = 10^3$ we do observe that infidelities go to below 10^{-4} for adequately long t_{end}. More interesting are the results for $\Omega_0 = 10^3\Gamma$. Here we obtain infidelities below 10^{-4} for any number of atoms we have tested except $N = 10^6$. Here we even see infidelities as low as 10^{-8}. We note that the two data-series for $N = 10^3$ and $N = 10^4$ coincide. This is because in this regime, the STIRAP procedure is so efficient that virtually no excited state population occurs. The main contributor to decoherence in this case is incomplete STIRAP processes, where the condensate does not reach the exact target state. To verify this claim, we have reevaluated these two data-series with spontaneous emission turned off and obtained data that coincides with these two series.

4. Conclusion and outlook

We have here proposed a novel scheme for population transfer in ^{87}Rb BECs which allows for the implementation of arbitrary SU(2) rotations. We have done so using optical pulses and

STIRAP, whereby we successfully mitigate decoherence caused by spontaneous emission. This has been confirmed by our results which show 10^{-7} infidelities for arbitrary rotations that take even less than 100 ns.

Potential applications include quantum metrology schemes that rely on extremely accurate preparation of BECs. In quantum magnetometry for instance [10], a BEC is typically prepared in some coherent superposition of magnetic sublevels, whose Larmor precession reveals the strength of some applied magnetic field. Conventionally experimenters here use radiofrequency and microwave pulses to create these desired superpositions. Here our method offers a very high fidelity alternative, that have a number of practical advantages. First and foremost, there appears to be no intrinsic limit to the fidelities, but rather a trade-off between fidelity and speed. Thus, the fidelity gained may be orders of magnitude closer to unity. Secondly, we find that our method is extremely fast, allowing full rotations to occur in less than 100 ns. As a second application, one may consider the growing field of BEC quantum information processing [11, 12]. This has several advantages over conventional quantum computing. One of which is the recent discovery, that one may accurately know the quantum state of a BEC with a minimum of disturbance, using phase-contrast imaging [13]. There are also some works which show that certain interactions between BECs are enhanced by the number of particles present, and schemes to produce useful entanglement have been studied theoretically [14]. As already discussed here, this bosonic enhancement is also a drawback, as it produces increased decoherence rates, but we have successfully demonstrated here that they are no longer an issue for single qubit gates based on STIRAP. As our scheme allows us to use optical pulses, the lasers may be tightly focused thus allowing several BECs to exist in close vicinity to each other. This enables scalable configurations where each spinor BEC can be addressed individually on e.g. an atom chip or similar device [8].

Acknowledgements

This work is supported by the Shanghai Research Challenge Fund, and National Natural Science Foundation of China grant 61571301, the Thousand Talents Program for Distinguished Young Scholars, and the New York University Global Seed Grants for Collaborative Research.

References

[1] Issouffa Y, Messikh A, Wahiddin M, Umarov B and Gharib M 2013 *Information and Communication Technology for the Muslim World (ICT4M)* pp 1–6
[2] Kis Z and Renzoni F 2002 *Phys. Rev. A* **65**(3) 032318
[3] Ivanov P A, Vitanov N V and Bergmann K 2005 *Phys. Rev. A* **72**(5) 053412
[4] Görlitz A, Chikkatur A and Ketterle W 2001 *Physical Review A* **63** 041601
[5] Petrosyan D and Mølmer K 2013 *Physical Review A* **87** 033416
[6] Møller D, Madsen L B and Mølmer K 2007 *Physical Review A* **75** 062302
[7] Kis Z and Stenholm S 2002 *Journal of Modern Optics* **49** 111–124
[8] Pyrkov A N and Byrnes T 2013 *New Journal of Physics* **15** 093019
[9] Steck D A 2001 Rubidium 87 d line data
[10] Vengalattore M, Higbie J, Leslie S, Guzman J, Sadler L and Stamper-Kurn D 2007 *Physical review letters* **98** 200801
[11] Byrnes T, Wen K and Yamamoto Y 2012 *Phys. Rev. A* **85**(4) 040306
[12] Byrnes T 2012 *World Acad. Sci. Eng. Technol* **63** 542
[13] Ilo-Okeke E O and Byrnes T 2014 *Physical review letters* **112** 233602
[14] Idlas S, Domenzain L, Spreeuw R and Byrnes T 2016 *Physical Review A* **93** 022319

Plasmonic solitons and dromions via plasmon-induced transparency

Zhengyang Bai[1] **and Guoxiang Huang**[1,2,†]
[1]State Key Laboratory of Precision Spectroscopy, School of Physical and Material Sciences, East China Normal University, Shanghai 200062, China
[2]NYU-ECNU Joint Institute of Physics at NYU-Shanghai, Shanghai 200062, China
E-mail: gxhuang@phy.ecnu.edu.cn

Abstract. We propose a method to enhance Kerr nonlinearities and realize low-power gigahertz solitons via plasmon induced transparency (PIT) in a new type of metamaterial, which is constructed by an array of unit cell consisting of a cut-wire and a pair of varactor-loaded split-ring resonators. We show that the PIT in such metamaterial can not only mimic the electromagnetically induced transparency in coherent three-level atomic systems, but also exhibit a crossover from PIT to Autler-Townes splitting. We further show that the system suggested here also possess giant second- and third-order nonlinear optical susceptibilities, which may be used to create plasmon solitons and dromions with extremely low power. Our studies raise the possibility for obtaining strong nonlinear effects of gigahertz radiation at very low intensity based on room temperature metamaterials.

1. Introduction

Recently, considerable attention has been paid to the investigation on plasmon-induced transparency (PIT) [1, 2, 3, 4], an interesting destructive interference effect arising from the strong coupling between wide-band bright oscillatory mode and narrow-band dark oscillatory mode of the meta-atoms in plasmonic metamaterials. PIT, a plasmonic analog of the atomic electromagnetically induced transparency (EIT) [5], can not only be used to significantly suppress the radiative loss of metamaterials, but also bring steep dispersion and hence large reduction of the propagation velocity of plasmonic polaritons.

Because of highly resonant (and hence dispersive) character inherent in PIT metamaterials, linear plasmon polaritons inevitably undergoes a significant distortion during propagation. Furthermore, due to diffraction effect, which can not be neglected for the cases of small transverse size or long propagation distance, a large deformation of linear plasmon polaritons is unavoidable. Thus it is necessary to extend the PIT to nonlinear propagation regime[6, 7, 8, 9].

Here, we give a report of our recent studies [10, 11, 12] on this subject. We elucidate that the PIT in such metamaterial can not only mimic the EIT in coherent three-level atomic systems, but also exhibit a crossover from PIT to Autler-Townes splitting. We show that the system suggested here also possess giant second-order and third-order nonlinear optical susceptibilities, which may be used to create plasmon solitons and dromions with extremely low generation power. Our studies raise the possibility for obtaining nonlinear plasmon polaritons at very low intensity based on room-temperature metamaterials.

2. Model

The metamaterial structure considered here is an array of PIT unit cell [13] consisting of a CW and two SRRs with a nonlinear varactor inserted into the slits of the SRRs [see Fig. 1(a) and Fig. 1(c)]. We assume that an incident gigahertz radiation is collimated on the metamaterial

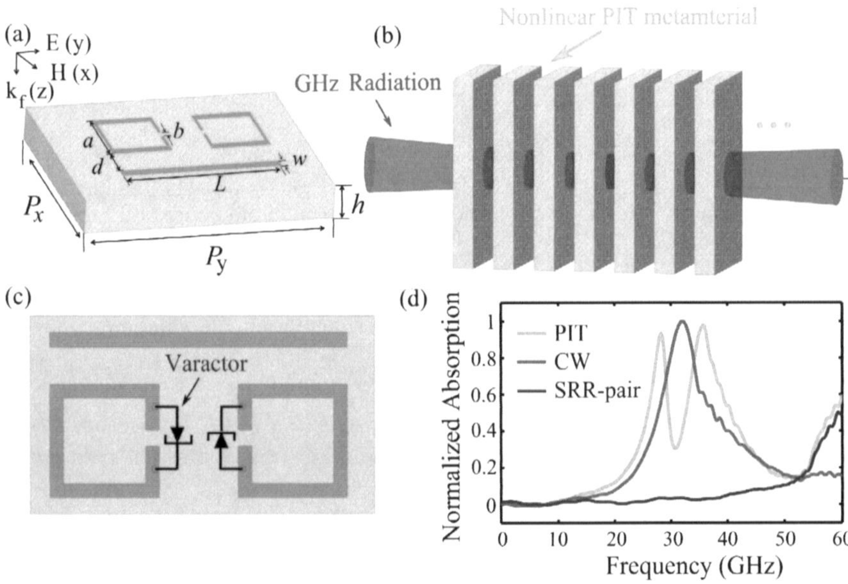

Figure 1. (a) Schematic of the unit cell (meta-atom) of the plasmonic metamaterial, which consists of a CW and a pair of SRRs. Geometrical parameters of the unit cell are $L = 1.7$, $w = 0.1$, $a = 0.58$, $b = 0.1$, $P_x = 1.6$, $P_y = 2.4$ (in unit mm). (b) SRR pair with a hyperabrupt tuning varactor mounted onto slits. (c) Possible experimental arrangement for the measurement of GHz radiation in the nonlinear PIT metamaterial. (d) Normalized absorption spectrum of the CW (red), the SRR-pair (blue) and the unit cell of PIT metamaterial (green).

array with the electric field parallel to the CW, as illustrated in Fig. 1(b) [11, 12, 13]. Normalized absorption spectrum of the sole-CW (red), SRR-pair (blue) and the unit cell of PIT metamaterial (green) are shown in Fig. 1(d). The CW array shows a typical localized surface plasmon resonance, while the SRRs support an inductive-capacitive (LC) resonance at the same frequency. The CW is directly excited by the incident electric field along the CW, while the SRRs are weakly coupled to the incident field due to the perpendicular orientation of the field. The near field coupling between the CW and SRRs excites the LC resonance in the SRRs, and hence the CW and SRRs serve respectively as the bright and dark modes in such excitation scheme, which leads to a dip at the center of the broad peak for the absorption spectrum.

The dynamics of the bright and dark modes in the unit cell at the position $\mathbf{r} = (x, y, z)$ can be described by the Lorentz equations for two coupled oscillators [1, 13]

$$\ddot{q}_1 + \gamma_1\dot{q}_1 + \omega_0^2 q_1 - \kappa^2 q_2 = g\,E(\mathbf{r}, t), \tag{1}$$
$$\ddot{q}_2 + \gamma_2\dot{q}_2 + (\omega_0 + \Delta)^2 q_2 - \kappa^2 q_1 + \alpha q_2^2 + \beta q_2^3 = 0, \tag{2}$$

where q_1 and q_2 are respectively amplitudes of the bright and dark modes (the dot over q_j denotes time derivative), with γ_1 and γ_2 respectively their damping rates; $\omega_0 = 2\pi$ 32 GHz and $\omega_0 + \Delta$

are respectively linear natural frequencies of the bright and dark modes ($\gamma_2 \ll \gamma_1 \ll \omega_0$); parameter κ denotes the coupling strength between the CW and SRR-pair; g is the parameter indicating the coupling strength of the bright mode with the incident radiation E. The last two terms on the left hand of Eq. (2) are provided by the hyperabrupt tuning varactors mounted onto gaps of the SRRs [7, 9]. Thus the metamaterial structure suggested here is a coupled anharmonic oscillator system with the nonlinear coefficients α and β and driven by the electric field E.

The equation of motion for the electric field E is governed by the Maxwell equation

$$\nabla^2 E - \frac{1}{c^2}\frac{\partial^2 E}{\partial t^2} = \frac{1}{\varepsilon_0 c^2}\frac{\partial^2 P}{\partial t^2}, \tag{3}$$

with the electric polarization intensity given by $P = \varepsilon_0 \chi_D^{(1)} E + Neq_1$, where N the density of unit cells, e is the unit charge, and $\chi_D^{(1)}$ is the optical susceptibility of the background material (which is assumed to be liner). In terms of the relation $P = \varepsilon_0 \chi E$, the electric susceptibility χ can be obtained by the formula

$$\chi = \chi_D^{(1)} + \frac{Ne}{\varepsilon_0 E} q_1. \tag{4}$$

We assume the incident radiation has frequency ω_f, which is near ω_0. Thus there is resonant interaction between the electric field E and the oscillators q_1 and q_2. To treat such resonant, nonlinear problem analytically, we assume $q_j = q_{dj} + (q_{fj} e^{i(k_0 z - \omega_0 t)} + \text{c.c.}) + (q_{sj} e^{2i(k_0 z - \omega_0 t)} + \text{c.c.})$, $E = E_d + (E_f e^{i(k_f z - \omega_f t)} + \text{c.c.}) + (E_s e^{i[(2k_f + \Delta k)z - 2\omega_f t]} + \text{c.c.})$. Here q_{dj}, q_{fj}, and q_{sj} are respectively amplitudes of the longwave (rectification field or mean field), shortwave (fundamental wave), and second harmonic wave of the jth oscillator, with k_0 (ω_0) the wavenumber (frequency) of the fundamental wave; E_d, E_f, and E_s are respectively amplitudes of the longwave, shortwave, and second harmonic wave of the electric field; k_f (ω_f) is the wavenumber (frequency) of the fundamental wave, and Δk is a detuning. From the Maxwell-Lorentz (ML) equations (1) and (3) and using rotating-wave and slowly varying envelope approximations, we can obtain a series of equations for the motion of $q_{\mu j}$ and E_μ ($\mu = d, f$)[11, 12].

In panels (a), (b), (c) of Fig. 2 we show respectively the numerical result of normalized absorption spectrum of the system versus frequency for separation $d = 0.38$, 0.24, and 0.02 (in unit mm), obtained by using the commercial finite difference time domain soft-ware package (CST Microwave Studio). We observe that a PIT transparency window in the absorption spectrum opens; furthermore, the transparency window becomes wider and deeper as d is reduced. These phenomena are clear indications of PIT-ATS crossover and will be analyzed in detail below.

3. PIT-ATS crossover

We now solve the equations for $q_{\alpha j}$ and E_α by using the method of multiple scales. Take the asymptotic expansion $q_{fj} = \epsilon q_{fj}^{(1)} + \epsilon^2 q_{fj}^{(2)} + \cdots$, $q_{dj} = \epsilon^2 q_{dj}^{(2)} + \cdots$, $q_{sj} = \epsilon^2 q_{sj}^{(2)} + \cdots$, $E_f = \epsilon E_f^{(1)} + \epsilon^2 E_f^{(2)} + \cdots$, $E_d = \epsilon^2 E_d^{(2)} + \cdots$, where ϵ is a dimensionless small parameter characterizing the amplitude of the incident electric field. All quantities on the right-hand side of the expansion are assumed as functions of the multi-scale variables $x_1 = \epsilon x$, $y_1 = \epsilon y$, $z_j = \epsilon^j z$ ($j = 0, 1, 2$) and $t_j = \epsilon^j t$ ($j = 0, 1$). Substituting this expansion into the the equations for $q_{\mu j}$ and E_μ and comparing the expansion parameter of each power ϵ, we obtain a chain of linear but inhomogeneous equations, which can be solved order by order.

At the first-order we obtain the solution for the shortwave field $E_f^{(1)} = F\exp[i(Kz_0 - \delta t_0)]$ where F is a yet to be determined envelope function depending on the "slow" variables x_1, y_1,

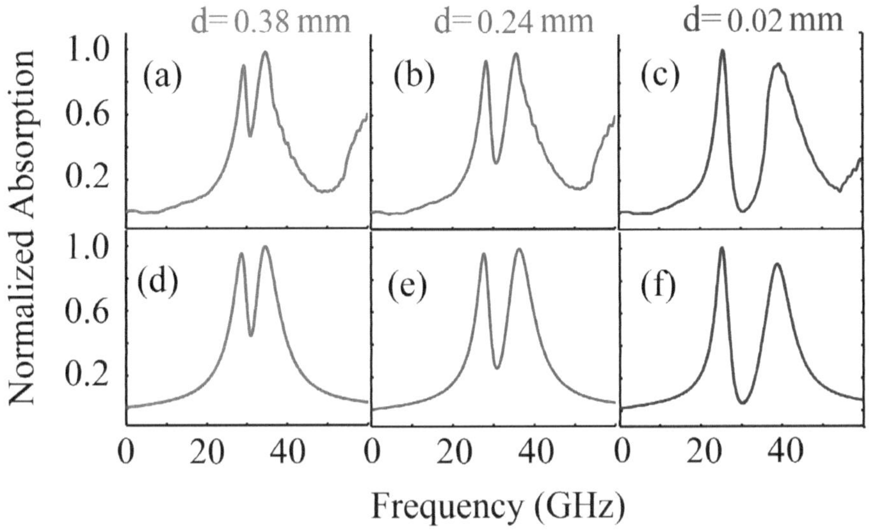

Figure 2. Numerical results of the normalized absorption spectrum of the metamaterial (Fig. 1) for (a) $d = 0.38$, (b) $d = 0.24$, and (c) $d = 0.02$ (in unit mm), respectively. Analytical results given in (d), (e), and (f) are obtained from solving the model Eqs. (1) and (3) in linear regime.

z_1, z_2, and t_1, $\delta = \omega_f - \omega_0$ is frequency detuning, and K is the linear dispersion relation given by

$$K = \frac{n_D}{c}\delta + \frac{\kappa_0 g D_2(\delta)}{D_1(\delta)D_2(\delta) - \kappa^4}. \tag{5}$$

Here $D_j(l\delta) = \omega_0^2 - l^2(\omega_0 + \delta)^2 - il\gamma_j(\omega_0 + \delta)$ $(j, l = 1, 2)$ and $\kappa_0 = (Ne\omega_0)/(2\varepsilon_0 c n_D)$.

Shown in panels (d), (e), (f) of Fig. 2 is the absorption spectrum Im(K) (the imaginary part of K) as a function of frequency. When plotting the figure we used the damping rates $\gamma_1 \approx 60$ GHz and $\gamma_2 \approx 10$ GHz, which are nearly independent of d, whereas κ decreases from 152.5 GHz at $d = 0.02$ mm to 69 GHz at $d = 0.38$ mm. We see that the analytical result (the lower part of Fig. 2) fits well with the numerical one (the upper part of Fig. 2).

To obtain a clear physical explanation on the variation of the PIT transparency window for different d shown in Fig. 2, we make a detailed analysis on the linear dispersion relation. Near the resonance point (i.e. $\delta \ll \omega_0$), we have $\omega_0^2 - (\omega_0 + \delta)^2 \approx -2\omega_0\delta$, and hence

$$K = \frac{n_D}{c}\delta + \frac{A(\delta + i\frac{\gamma_2}{2})}{B^2 - (\delta + i\frac{\gamma_1}{2})(\delta + i\frac{\gamma_2}{2})}, \tag{6}$$

with $A = \kappa_0 g/(2\omega_0)$ and $B = \kappa^2/(2\omega_0)$.

The linear dispersion relation (6) is similar to that obtained in Λ-type three-level atomic systems [5]. The PIT condition of the system reads $B^2 \gg \gamma_1\gamma_2/4$. To illustrate the interference effect between the bright and dark modes apparently, we make a spectrum decomposition on Im(K) using the technique developed recently in Refs. [14, 15] for different coupling constant κ.

(i). *Weak coupling region* $(\kappa^2 < (\gamma_1 - \gamma_2)\omega_0/2)$: In this region, the absorption spectrum can be expressed as

$$\mathrm{Im}(K) = -\frac{AC_+W_+}{\delta^2 + W_+^2} - \frac{AC_-W_-}{\delta^2 + W_-^2}, \tag{7}$$

where $W_\pm = -(\gamma_1+\gamma_2)/4 \pm [(\gamma_1-\gamma_2)^2/16 - B^2]^{\frac{1}{2}}$ and $C_\pm = \pm(\gamma_2/2 + W_\pm)/[(\gamma_1-\gamma_2)^2/4 - 4B^2]^{\frac{1}{2}}$. We see that Im$(K)$ consists of two Lorentzians terms. Fig. 3(a) shows Im(K) as a function of δ

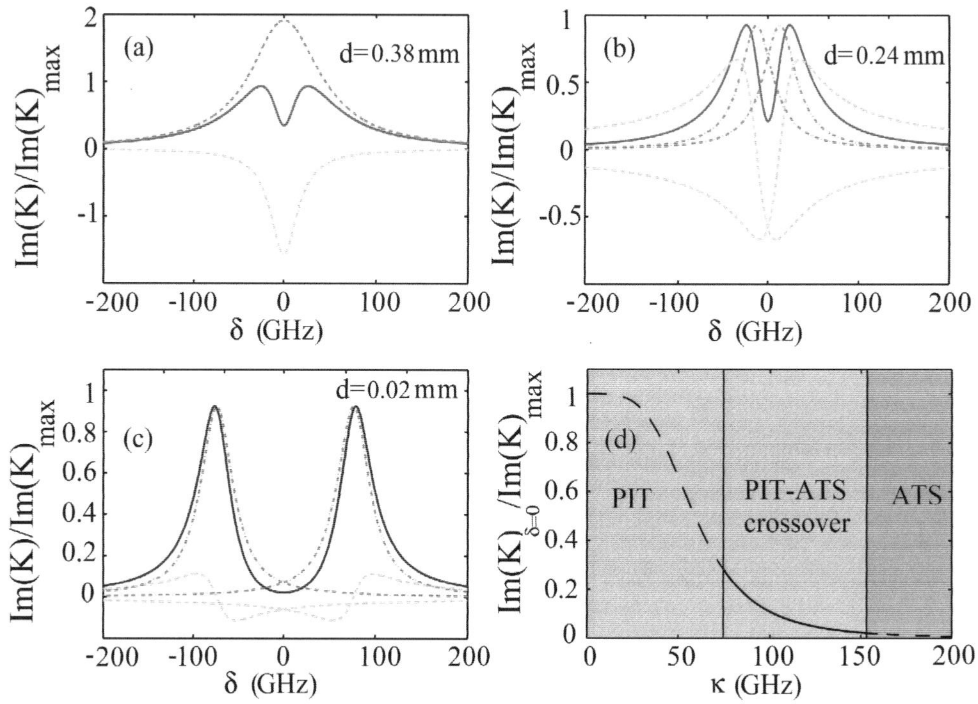

Figure 3. (a) The dashed (dotted-dashed) line is the result of the first (second) Lorentzian term in Eq. (7). The solid line is the sum of the two Lorentzian terms, giving the absorption spectrum Im(K) in the weak coupling (PIT) region (d = 0.38 mm). (b) The dashed-dotted lines denote the first two Lorentzian terms in Eq. (8); the dashed lines denote the third and fourth terms in Eq. (8). The solid line is the sum of the all four terms, giving Im(K) in the intermediate coupling (PIT-ATS crossover) region (d = 0.24 mm). (c). The same as (b) but with $d = 0.02$ mm, giving Im(K) in the strong coupling (ATS) region. (d) Transition from PIT to ATS when κ changes.

for $\kappa = 69$ GHz. The dashed (dotted-dashed) line is the result of the first (second) Lorentzian term in Eq. (7), which is negative (positive). Because the two Lorentzians terms have the same center position but opposite sign, their superposition gives a destructive interference between CW and the SRR-pair. As a result, a small dip in Im(K) curve (i.e. the solid line) appears. Such phenomenon belongs PIT in nature.

(ii). *Intermediate coupling region* ($\kappa^2 > (\gamma_1 - \gamma_2)\omega_0/2)$): In this case, we obtain

$$\begin{aligned}\mathrm{Im}(K) = & \frac{A(\gamma_1+\gamma_2)/8}{(\delta-\delta_0)^2+(\gamma_1+\gamma_2)^2/16}+\frac{A(\gamma_1+\gamma_2)/8}{(\delta+\delta_0)^2+(\gamma_1+\gamma_2)^2/16} \\ & -\frac{Ah(\delta-\delta_0)}{(\delta-\delta_0)^2+(\gamma_1+\gamma_2)^2/16}+\frac{Ah(\delta+\delta_0)}{(\delta+\delta_0)^2+(\gamma_1+\gamma_2)^2/16},\end{aligned} \tag{8}$$

with $\delta_0 = \sqrt{B^2 - (\gamma_1 - \gamma_2)^2/16}$ and $h = (\gamma_2 - \gamma_1)/(8\delta_0)$. We see that the absorption spectrum is made of four terms. The first two terms are Lorentzians with the same width but different center position, which are two resonance peaks belonging respectively to the CW and the SRR-pair. The dip between the two Lorentzians can be interpreted as a gap between two resonances, which is a typical character of ATS. The next two terms are interference terms. Because they lowers the dip formed by the first two terms, a destructive interference (i.e. PIT) occurs. Since both PIT and ATS occur simultaneously in the system, we assign such phenomenon as *PIT-ATS crossover*. Shown in Fig. 3(b) is the result of various terms in Eq. (8) and the total absorption spectrum Im(K) for $\kappa = 82$ GHz.

(iii). *Strong coupling region* ($\kappa^2 \gg (\gamma_1 - \gamma_2)\omega_0/2)$): The spectrum decomposition of Im(K) is still given by Eq. (8), but the destructive interference effect contributed by the last two terms plays a negligible role. Fig. 3(c) shows the result for $\kappa = 152.5$ GHz. We see that in the strong coupling region the absorption spectrum of the electric field displays mainly the character of ATS.

In Fig. 3(d) we show the "phase diagram" of the system, where three regions (i.e. PIT, PIT-ATS crossover, and ATS) are indicated clearly. We see that a transition from PIT to ATS indeed occurs when κ changes from small to large values.

4. Enhanced Kerr effect

We now extend the equations for $q_{\mu j}$ and E_μ to higher order to demonstrate enhanced Kerr effect due to the resonant interaction.

At the second-order, a divergence-free condition requires $\partial F/\partial z_1 + (1/V_g)\partial F/\partial t_1 = 0$, where $V_g = (\partial K/\partial \delta)^{-1}$ is the group velocity of the shortwave envelope F. The solution for the longwave (rectification) field reads $E_d^{(2)} = G$. With the above results we proceed to the third order. The solvability condition at this order yields the nonlinear equation

$$i\frac{\partial F}{\partial z_2} - \frac{1}{2}K_2\frac{\partial^2 F}{\partial \tau_1^2} + \frac{c}{2\omega_0 n_D}\left(\frac{\partial^2}{\partial x_1^2} + \frac{\partial^2}{\partial y_1^2}\right)F + \frac{\omega_0}{2cn_D}\chi^{(3)}|F|^2 F e^{-2\bar{\alpha} z_2} + \frac{m_1\omega_0}{2cn_D}\chi^{(2)}GF = 0. \quad (9)$$

Here $\tau_1 = \epsilon\tau$ ($\tau \equiv t - z/V_g$), $\bar{\alpha} = \epsilon^{-2}\mathrm{Im}(K)$ is the coefficient describing linear absorption, $K_2 = \partial^2 K/\partial\delta^2$ is the coefficient describing group-velocity dispersion, $m_1 \equiv |D_1(\delta)D_2(\delta) - \kappa^4|^2/(D_1(\delta)D_2(\delta) - \kappa^4)^2$, $\chi^{(2)}$ and $\chi^{(3)}$ are respectively the second-order and third-order nonlinear susceptibilities with the form

$$\chi^{(2)} = \frac{-2Neg^2\kappa^6\alpha}{\epsilon_0(\omega_0^4 - \kappa^4)|D_1(\delta)D_2(\delta) - \kappa^4|^2}, \quad (10)$$

$$\chi^{(3)} = \left(\frac{4\alpha^2\omega_0^2}{\omega_0^4 - \kappa^4} + \frac{2\alpha^2 D_1(2\delta)}{D_1(2\delta)D_2(2\delta) - \kappa^4} - 3\beta\right) \times \frac{g^3\kappa^8 Ne}{\epsilon_0(D_1(\delta)D_2(\delta) - \kappa^4)^2|D_1(\delta)D_2(\delta) - \kappa^4|^2}. \quad (11)$$

We see that $\chi^{(2)}$ is proportional to the parameter α, i.e. it is contributed by the quadratic nonlinearity in Eq. (2); $\chi^{(3)}$ is proportional to the parameters α and β, which means that it comes from the contributions by the quadratic and cubic nonlinearities in Eq. (2).

Furthermore, the nonlinear equation for the longwave (rectification) field G arises at the fourth-order approximation, which reads

$$\left(\frac{\partial^2}{\partial x_1^2} + \frac{\partial^2}{\partial y_1^2}\right)G - \left(\frac{1}{V_p^2} - \frac{1}{V_g^2}\right)\frac{\partial^2 G}{\partial \tau_1^2} - \frac{\chi^{(2)}}{c^2}\frac{\partial^2|F|^2}{\partial \tau_1^2}e^{-2\bar{\alpha} z_2} = 0, \quad (12)$$

where V_p is the phase velocity of the longwave field G, defined by $1/V_p^2 = (n_D/c)^2 + (Neg\omega_0^2)/[\varepsilon_0 c^2(\omega_0^4 - \kappa^4)]$. The last term on the left hand of Eq. (12) corresponds to a second-order plasmonic rectification. Above results tell us that the self-interaction of the shortwave

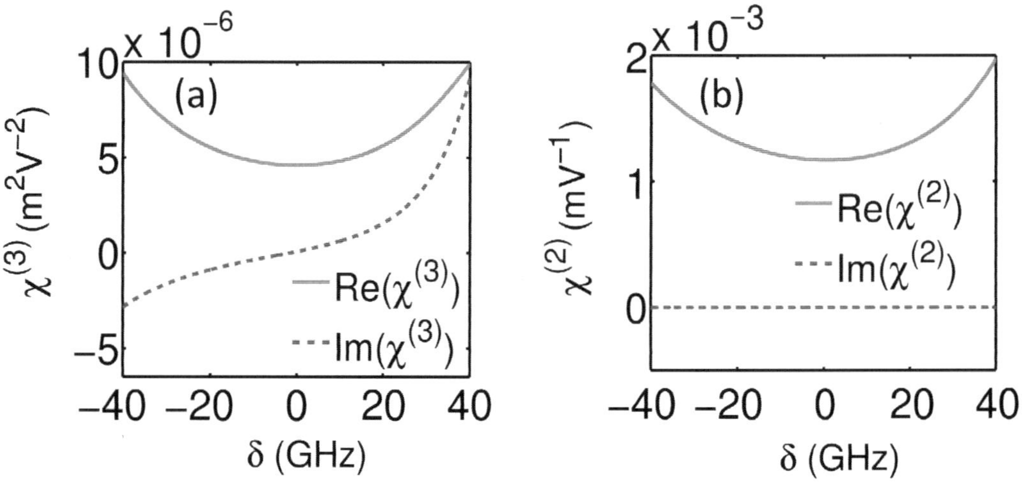

Figure 4. Nonlinear susceptibilities of the PIT metamaterial. (a) Real and imaginary parts of the third-order susceptibility $\chi^{(3)}$ [i.e. $\mathrm{Re}(\chi^{(3)})$ and $\mathrm{Im}(\chi^{(3)})$] as functions of the frequency detuning δ. (b) Real and imaginary parts of the second-order susceptibility $\chi^{(2)}$ [i.e. $\mathrm{Re}(\chi^{(2)})$ and $\mathrm{Im}(\chi^{(2)})$] as functions of δ. System parameters used are given in the text.

(with the envelope F) can stimulate the generation of the longwave field G [Eq. (12)], and at the same time longwave field G has a back-action to the shortwave field F [Eq. (9)]. Equations (9) and (12) are the general form of Davey-Stewartson (DS) equations describing the propagation of high-dimensional nonlinear plasmonic polaritons in the system.

In Fig. 4(a) and Fig. 4(b) we show respectively curves of $\chi^{(3)}$ and $\chi^{(2)}$ as functions of the frequency detuning δ. We observe that $\chi^{(2)}$ is nearly real and has the order of magnitude 10^{-3} mV^{-1}; The real part of the third-order susceptibility, $\mathrm{Re}(\chi^{(3)})$, has the order of magnitude 10^{-6} $\mathrm{m}^2\mathrm{V}^{-2}$. The physical reason for such large second- and third-order nonlinearities predicted here is due to the fact that the incident radiation E is resonant with the oscillators q_1, q_2 and the system works under PIT condition. We also show the imaginary part of $\chi^{(3)}$, which contributes a nonlinear absorption to the radiation field, is much less than the real part $\mathrm{Re}(\chi^{(3)})$ when the system works in PIT transparency window. Such suppression of the nonlinear absorption is also due to the PIT effect.

Interestingly, the third-order nonlinear susceptibility $\chi^{(3)}$ can be further enhanced by using longwave-shortwave resonance. For simplicity, we assume that the transverse spatial distribution of the radiation is large enough so that the diffraction effect (i.e. its dependence on the transverse coordinates x and y) of the system can be neglected. Then from Eq. (12) we obtain $G = \chi^{(2)}|E_f|^2/[c^2(1/V_g^2 - 1/V_p^2)]$. Plugging this result into Eq. (9), we obtain an effective third-order nonlinear susceptibility

$$\chi_{\mathrm{eff}}^{(3)} = \chi^{(3)} + \chi_{\mathrm{SL}}^{(3)}, \tag{13}$$

$$\chi_{\mathrm{SL}}^{(3)} = m_1 \frac{\left(\chi^{(2)}\right)^2}{c^2\left(\frac{1}{V_p^2} - \frac{1}{V_g^2}\right)}, \tag{14}$$

where the subscript "SL" means that the corresponding term is due to the longwave-shortwave interaction. Equation (13) tells us that if some region of system parameters can be found where $V_p \approx V_g$, in addition to the PIT enhancement, the effective third-order nonlinear susceptibility

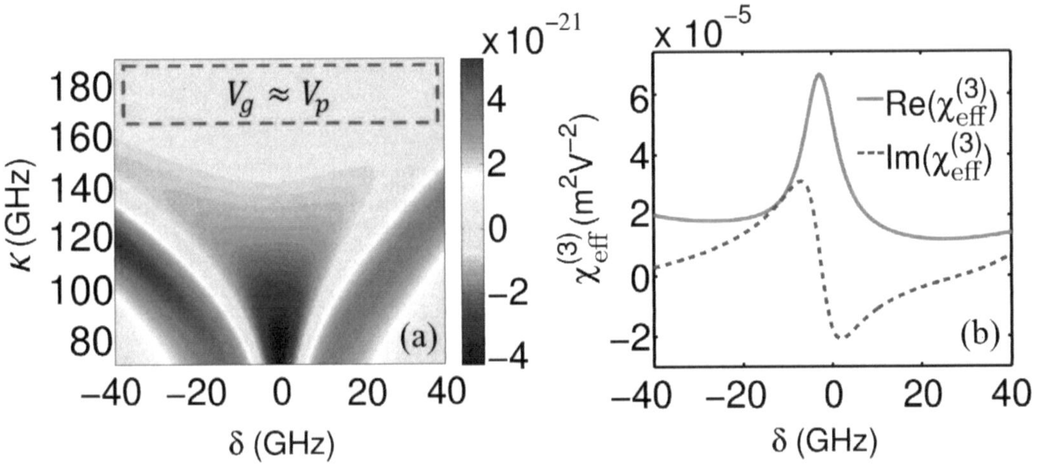

Figure 5. (a) The denominator $1/V_p^2 - 1/V_g^2$ of $\chi_{\rm SL}^{(3)}$ as functions of frequency detuning δ and the coupling coefficient κ. The rectangle enclosed by purple dashed lines shows the region where the longwave-shortwave resonance occurs (i.e. $V_g \approx V_p$). (b) Real part $\mathrm{Re}(\chi_{\rm eff}^{(3)})$ (orange solid line) and imaginary part $\mathrm{Im}(\chi_{\rm eff}^{(3)})$ (green dashed line) of the effective third-order nonlinear susceptibility $\chi_{\rm eff}^{(3)}$ as functions of frequency detuning δ for $\kappa = 180$ GHz.

$\chi_{\rm eff}^{(3)}$ can be further enhanced because of the drastic enhancement of $\chi_{\rm SL}^{(3)}$.

In the situation $V_p \approx V_g$, the system undergoes a resonant interaction between the longwave G and the shortwave F, a special form of three-wave resonance satisfying the conditions $\omega_1 + \omega_2 = \omega_3$ and $k_f(\omega_1) + k_f(\omega_2) = k_f(\omega_3)$. This point can be illustrated clearly if we choose $\omega_1 = \omega - \varepsilon\delta$, $\omega_2 = 2\varepsilon\delta$, and $\omega_3 = \omega + \varepsilon\delta$. Then one has $\omega_1 + \omega_2 = \omega_3$, and $k_f(\omega_1) + k_f(\omega_2) - k_f(\omega_3) = k_f(\omega - \varepsilon\delta) + k_f(2\varepsilon\delta) - k_f(\omega + \varepsilon\delta) = -2\varepsilon\delta\, \partial k_f/\partial\omega + k_f(2\varepsilon\delta) + O(\varepsilon^3)$ which is zero to order ε^3 if $\partial k_f/\partial\omega \simeq k_f(2\varepsilon\delta)/2\varepsilon\delta$, i.e. $V_g \simeq V_p$ [16].

In our system, a broad parameter region for $V_p \approx V_g$ exists. Illustrated in Fig. 5(a) is the denominator $1/V_p^2 - 1/V_g^2$ of $\chi_{\rm SL}^{(3)}$ as functions of the frequency detuning δ and the coupling coefficient κ. The rectangle enclosed by purple dashed lines in upper part of the figure illustrates the parameter region where $V_g \approx V_p$. Consequently, the longwave-shortwave resonance can indeed occur in the present PIT metamaterial, and based on this the effective third-order nonlinear susceptibility $\chi_{\rm eff}^{(3)}$ of the system can be enhanced greatly.

Fig. 5(b) shows curves of the real part and the imaginary part of the effective third-order nonlinear susceptibility $\chi_{\rm eff}^{(3)}$ as functions of δ for $\kappa = 180$ GHz. We observe that $\mathrm{Re}(\chi_{\rm eff}^{(3)})$ (which is much larger $\mathrm{Im}(\chi_{\rm eff}^{(3)})$ near $\delta = 0$) is enhanced one order of magnitude (up to the value 6.64×10^{-5} m^{2}V^{-2}), which is contributed by the longwave-shortwave resonance interaction. When plotting the figure, other system parameters used are $\gamma_1 = 60$ GHz, $\gamma_2 = 10$ GHz, $g = 1.79 \times 10^{11}$ C$\cdot$kg^{-1}, $\kappa_0 = 2.8 \times 10^{-8}$ kg$\cdot$C^{-1}cm^{-1}GHz^2, obtained by fitting the numerical result given in Fig. 1(d); parameters $\alpha = -1.27 \times 10^{15}$ cm^{-1}GHz^2 and $\beta = 2.26 \times 10^{25}$ cm^{-2}GHz^2 are derived from the result given in Ref. [7].

5. Plasmonic solitons and dromions

5.1. Plasmonic solitons

In this section, we focus on the regime that transverse spatial distribution of the radiation is large enough so that the diffraction effect (i.e. its dependence on the transverse coordinates x and y) of the system can be neglected. Based on such condition and returning to the original variables, Eq. (9) returns to the nonlinear Schrödinger (NLS) equation

$$i\left(\frac{\partial}{\partial z}+\alpha_1\right)U-\frac{1}{2}K_2\frac{\partial^2 U}{\partial \tau^2}-W|U|^2U=0, \tag{15}$$

where $W=-[\omega_0/(2cn_D)]\chi^{(3)}$, $\tau=t-z/V_g$ and $U=\epsilon F\exp(-\alpha_1 z)$.

Generally, Eq. (15) has complex coefficients and hence it is a Ginzburg-Landau equation. However, due to the PIT effect the imaginary part of the complex coefficients can be made much smaller than their real part. Thus Eq. (15) can be converted into the dimensionless form

$$i\frac{\partial}{\partial s}u+\frac{1}{2}\frac{\partial^2 u}{\partial \sigma^2}+|u|^2u=id_0u, \tag{16}$$

with $z=-L_{\rm Disp}s$, $\tau=\tau_0\sigma$, $U=U_0u$, and $d_0=L_{\rm Disp}/L_A$. Here τ_0, $L_{\rm Disp}=\tau_0^2/\widetilde{K}_2$, $L_A=1/\alpha_1$, and $U_0=(1/\tau_0)\sqrt{\widetilde{K}_2/\widetilde{W}}$ are typical pulse length, dispersion length, absorption length, and amplitude of the pulse, respectively (the tilde above the quantity means taking real part). Notice that for forming solitons $L_{\rm Disp}$ has been set to equal typical nonlinearity length $L_{\rm Nonl}\,(\equiv 1/(U_0^2\widetilde{W}))$.

If d_0 is small, the term d_0u can be taken as a perturbation. One can use the perturbation method for solitons to solve the Eq. (16) to obtain a single-soliton solution under the perturbation. The result is given by $u=\eta e^{d_0 s}\,{\rm sech}[2\eta\, e^{d_0 s}\,(\sigma-\sigma_0+4\varsigma s)]e^{-2i\varsigma\sigma-4i(\varsigma^2-\eta^2)s-i\phi_0}$, where η, ς, σ_0 and ϕ_0 are real free parameters determining the amplitude (as well as the width), propagating velocity, initial position, and initial phase of the soliton, respectively. When taking $\eta=1/2$, $\varsigma=\sigma_0=\phi_0=0$ and noting that $s=-z/L_{\rm Disp}$, we obtain the electric field corresponding the above single-soliton solution

$$E=\frac{1}{\tau_0}\sqrt{\frac{\widetilde{K}_2}{\widetilde{W}}}{\rm sech}\left[\frac{1}{\tau_0}\left(t-\frac{z}{\widetilde{V}_g}\right)e^{-z/L_A}\right]e^{i\Phi(z,t)-z/L_A}+{\rm c.c.} \tag{17}$$

with $\Phi(z,t)\equiv[\widetilde{K}+k_f-1/(L_D)]z-\omega_f t$, which describes a damped bright soliton traveling with velocity $\widetilde{V}_g$.

For $\delta=15$ GHz and with other parameters given above, we obtain numerical values of the coefficients in Eq. (15), given by $\alpha_1=0.0278\ {\rm cm}^{-1}$, $K_2=(5.02+0.57i)\times10^{-23}{\rm cm}^{-1}{\rm s}^2$, $W=(4.62+0.72i)\times10^{-2}\ {\rm C}^2{\rm kg}^{-2}{\rm cm}^{-3}{\rm s}^4$. We see that, as expected, the imaginary part of these coefficients are indeed much smaller than their real part. By taking $\tau_0=1.5\times10^{-11}$ s, we obtain $U_0=2.2\ {\rm V\,cm}^{-1}$, $L_{\rm Disp}=L_{\rm Nonl}=4.48$ cm, and $L_A=36$ cm. Thus one has $d_0=0.12$ and hence the dissipation in the system can be indeed taken as a perturbation.

The power associated with the soliton is given by Poynting's vector integrated over the cross-section of the radiation in the transverse directions, i.e. $P=\int(\mathbf{E}\times\mathbf{H})\cdot\mathbf{e}_z dS$, where $\mathbf{e}_z$ is the unit vector in the propagation direction. We obtain the peak power of the soliton $\bar{P}_{\rm max}=2\epsilon_0 cn_pS_0\widetilde{K}_2/(\widetilde{W}\tau_0^2)$, here $n_p=1+c\widetilde{K}/\omega_0$ is the refractive index and S_0 is the area of the intensity distribution of the radiation field in the transverse directions. Using the above

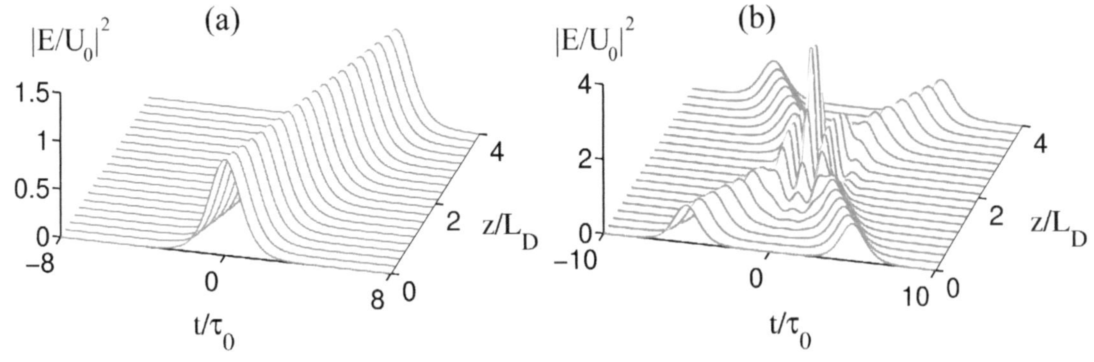

Figure 6. Propagation of the soliton and the interaction between two solitons. (a) The radiation intensity $|E/U_0|^2$ of the soliton as a function of t/τ_0 and z/L_D. (b) The collision between two solitons.

parameters and taking $S_0 = 6 \times 10^{-4}$ m^2, the peak power for generating the soliton is found to be $\bar{P}_{\rm max} = 568\,{\rm mW}$, which means that to generate the soliton in present system a very low input power is needed. This is a drastically contrast to the case in conventional media such as optical fibers, where pico- or femto-second laser pulses are needed to reach a very high peak power (usually at the order of several hundred kW) to stimulate enough nonlinearity for the formation of solitons.

The stability of the gigahertz soliton is tested by using numerical simulation. Fig. 6(a) shows the result of the radiation intensity $|E/U_0|^2$ of the soliton as a function of t/τ_0 and z/L_D. The solution is obtained by numerically solving Eq. (15) with the complex coefficients taken into account, with the initial condition given by $E(0,t)/U_0 = {\rm sech}(t/\tau_0)$. We see that the shape of the soliton undergoes no apparent deformation during propagation.

The collision between two solitons is also studied numerically, with the result shown in Fig. 6(b). The initial condition is given by $E(0,t)/U_0 = {\rm sech}[(t-3.0)/\tau_0]+1.2{\rm sech}[1.2(t+3.0)/\tau_0]$. We see that the both solitons can resume their original shapes after the collision, indicating that solitons in the PIT-metamaterial are robust during interaction.

5.2. Plasmonic dromions

We will take the diffraction effect into consideration in this section. Now we turn to the formation and propagation of high-dimensional nonlinear plasmon polaritons in the system by examining possible dromion solutions of the Eq. (9) and Eq. (12). After returning to original variables and converting them into dimensionless forms, Eq. (9) and Eq. (12) become

$$i\frac{\partial u}{\partial s} + \left(\frac{\partial^2}{\partial \xi^2} + g_y\frac{\partial^2}{\partial \eta^2} + g_{d1}\frac{\partial^2}{\partial \sigma^2}\right)u + 2g_1|u|^2u + g_2 vu = -id_0 u, \tag{18}$$

$$g_{d2}\frac{\partial^2 v}{\partial \sigma^2} - \left(\frac{\partial^2}{\partial \xi^2} + g_y\frac{\partial^2}{\partial \eta^2}\right)v + g_3\frac{\partial^2|u|^2}{\partial \sigma^2} = 0, \tag{19}$$

where $u = \epsilon F\exp(-\bar{\alpha}z_2)/U_0$, $v = \varepsilon^2 G/V_0$, $s = z/(2L_{\rm Diff})$, $\sigma = (t - z/\tilde{V}_g)/\tau_0$, $\xi = x/R_x$, $\eta = y/R_y$, $g_y = (R_x/R_y)^2$, $g_{d1} = L_{\rm Diff}/L_{\rm Disp}$, $g_{d2} = R_x^2(1/V_p^2 - 1/V_g^2)/\tau_0^2$, $g_1 = L_{\rm Diff}/L_{\rm Nonl}$, $g_2 = L_{\rm Diff}m_1\omega_0V_0\chi^{(2)}/(cn_D)$, $g_3 = \chi^{(2)}R_x^2U_0^2/(c^2\tau_0^2V_0)$, and $d_0 = 2L_{\rm Diff}/L_A$. Here R_x (R_y) is typical radius of the incident pulse in x (y) direction; U_0 and V_0 are respectively amplitudes of the shortwave envelope and the longwave; $L_{\rm Disp} = -\tau_0^2/\widetilde{K}_2$, $L_{\rm Diff} = \omega_0 n_D R_x^2/c$, $L_{\rm Nonl} =$

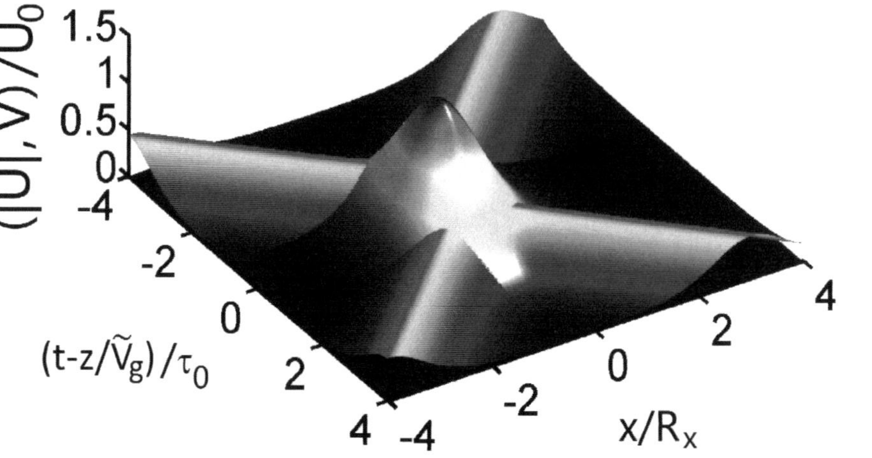

Figure 7. A single dromion excitation in the PIT metamaterial, which consists of a localized envelope for the shortwave component and two "tracks" for the longwave component. Plotted here is the shortwave envelope $|u| = |U|/U_0$ (the hump at the center) and the longwave (rectification) field $v = V/U_0$ [the two crossed tracks (plane solitons)] as functions of $\xi = x/R_x$ and $\sigma = (t - z/\tilde{V}_g)/\tau_0$. System parameters are chosen as $\mu = 1$, $\lambda = 1$, $a_1 = 0$, $p = 0$, $\varphi_\rho = 0$, $\varphi_\gamma = 0$ at $s = 0$.

$2cn_D/[\omega_0\tilde{\chi}^{(3)}U_0^2]$ and $L_A = 1/\mathrm{Im}(K)$ are respectively typical dispersion length, diffraction length, nonlinear length and absorption length. Note that when obtaining Eq. (18) we have neglected the imaginary parts of K_2 and $\chi^{(3)}$ under the PIT condition $\kappa^2 \gg \omega_0\sqrt{\gamma_1\gamma_2}$.

Expressions (18) and (19) are coupled (3+1)D nonlinear partial differential equations including effects of dispersion, diffraction, nonlinearity, and a small loss caused by Im(K). Because a general consideration to obtain stable high-dimensional nonlinear solutions of such equations is not available yet, here we consider only a specific case to seek possible (2+1)D dromion solutions by using some assumptions for simplification. First, we assume $L_{\rm Disp}$, $L_{\rm Diff}$ and $L_{\rm Nonl}$ have the same order of magnitude, which can be achieved by taking $\tau_0 = \sqrt{-\omega_0 n_D\tilde{K}_2/cR_x}$ and $U_0 = \sqrt{2c^2n_D/(\omega_0^2 n_D R_x^2\tilde{\chi}^{(3)})}$. Second, we choose realistic system parameters $\delta = -10$ GHz, $R_x = 0.47$ cm, $R_y = 2.5$ cm, $\tau_0 = 2.21 \times 10^{-11}$ s, $V_0 = 0.51\mathrm{V\,cm}^{-1}$, $\alpha = -3.99 \times 10^{14}$ cm^{-1}GHz^2, $\beta = 4.36 \times 10^{25}$ cm^{-2}GHz^2, $\kappa = 198$GHz. Then we obtain $L_{\rm Diff}$ $(= L_{\rm Disp} = L_{\rm Nonl}) = 4.99$ cm, d_0=0.064, $U_0 = 4.57$ Vcm^{-1}, and $g_{d1} = g_1 = g_2 = g_{d2} = 1$, $g_3 = 4$, $g_y \approx 0$. Because d_0 is small, the term on the right hand side of Eq. (18) can also be taken as a perturbation. As a first step, we neglect such perturbation and hence Eq. (18) and Eq. (19) are simplified into standard Davey-Stewartson-I (DSI) equations

$$i\frac{\partial u}{\partial s} + \frac{\partial^2 u}{\partial \sigma^2} + \frac{\partial^2 u}{\partial \xi^2} + 2|u|^2 u + vu = 0, \tag{20}$$

$$\frac{\partial^2 v}{\partial \sigma^2} - \frac{\partial^2 v}{\partial \xi^2} + 4\frac{\partial^2 |u|^2}{\partial \sigma^2} = 0. \tag{21}$$

which are completely integrable and can be solved exactly by the use of inverse scattering transform. One of remarkable properties of the DSI equations is that they allow various dromion solutions. A single dromion solution of the DSI equations reads $u = Q/P$, $v = 4\partial^2 \ln P/\partial\sigma^2$, where $P = 1 + \exp(\eta_1 + \eta_1^*) + \exp(\eta_2 + \eta_2^*) + \gamma\exp(\eta_1 + \eta_1^* + \eta_2 + \eta_2^*)$ and $Q = \rho\exp(\eta_1 + \eta_2)$, with

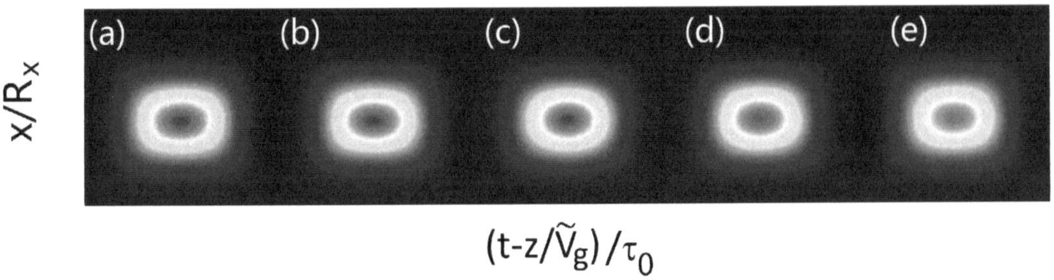

Figure 8. Spatiotemporal evolution of the plasmon dromion as a function of $\xi = x/R_x$ and $\sigma = (t - z/\tilde{V}_g)/\tau_0$ based on Eq. (9) and Eq. (12), by taking the solution (22) plus 10% random disturbance as an initial condition. (a) Initial profile of the dromion ($z = 0$). (b), (c), (d), and (e) are dromion profiles when propagating respectively to $z = 0.5\,L_{\rm Diff}$, $1.0\,L_{\rm Diff}$, $1.5\,L_{\rm Diff}$, $2.0\,L_{\rm Diff}$, with $L_{\rm Diff} = 4.99$ cm.

$\eta_1 = (k_r + ik_i)(\xi + \sigma)/\sqrt{2} + (\Omega_r + i\Omega_i)s$, $\eta_2 = (l_r + il_i)(\xi - \sigma)/\sqrt{2} + (\omega_r + i\omega_i)s$, $\Omega_r = -2k_r k_i$, $\omega_r = -2l_r l_i$, $\Omega_i + \omega_i = k_r^2 + l_r^2 - k_i^2 - l_i^2$, $\rho = |\rho|\exp(i\varphi_\rho)$, and $|\rho| = 2[2k_r l_r(\gamma - 1)]^{1/2}$. Here, k_r, k_i, l_r, l_i, $|\rho|$, φ_ρ, and γ are real integration constants. If we choose $k_r l_r > 0$, we have $\gamma = \exp(2\varphi_\gamma)$ with $\varphi_\gamma > 0$. By taking $k_r = \sqrt{2}\mu$, $k_i = \sqrt{2}a_1$, $l_r = \sqrt{2}\lambda$, $l_i = \sqrt{2}p$ ($\lambda\mu \geq 0$), $\Omega_i = 2(\mu^2 - a_1^2)$, $\omega_i = 2(\lambda^2 - p^2)$, $\Omega_r = -4a\mu$, and $\omega_r = -4\lambda p$, we obtain explicit expressions

$$u = \frac{2\mu\exp(ih)}{m\cosh f_1 + n\cosh f_2}, \tag{22a}$$

$$\begin{aligned} v = & \frac{4(m^2+n^2)(\mu^2+\lambda^2) - 8\mu^2}{(m\cosh f_1 + n\cosh f_2)^2} \\ & + \frac{8mn[(\mu^2+\lambda^2)\cosh f_1 \cosh f_2 - (\mu^2 - \lambda^2)\sinh f_1 \sinh f_2]}{(m\cosh f_1 + n\cosh f_2)^2}, \end{aligned} \tag{22b}$$

where $m = \{\mu/[\lambda(\gamma-1)]\}^{1/2}$, $n = \{\mu\gamma/[\lambda(\gamma-1)]\}^{1/2}$, $h = a_1(\sigma+\xi) + p(\sigma-\xi) + 2(\mu^2+\lambda^2-a_1^2-p^2)s + \varphi_\rho$, $f_1 = \mu(\sigma+\xi) - \lambda(\sigma-\xi) - 4(a_1\mu - \lambda p)s$, $f_2 = \mu(\sigma+\xi) + \lambda(\sigma-\xi) - 4(a_1\mu + \lambda p)s + \varphi_\gamma$. Obviously, the dromion given above consists of a localized envelope u for the shortwave component, which decays exponentially in all spatial directions (shown by the hump at the center of Fig. 7), and two plane solitons for the longwave component v, in which each plane soliton decays in its traveling direction (shown by the two "tracks" in Fig. 7).

The result presented above is the dromion solution based on Eq. (20) and Eq. (21) without considering the loss in the system. It is necessary to investigate the spatiotemporal evolution of the dromion and its stability starting directly from Eq. (9) and Eq. (12). We make a numerical simulation on Eq. (9) and Eq. (12) by taking the dromion solution (22) with a random disturbance as an initial condition. Shown in Fig. 8 is the evolution of the dromion pulse as a function of $\xi = x/R_x$ and $\sigma = (t - z/\tilde{V}_g)/\tau_0$ by taking $\varepsilon = 0.1$. The profiles from (a) to (e) in the figure are respectively for the propagation distance $z = 0, 0.5L_{\rm Diff}$, $L_{\rm diff}$, $1.5L_{\rm diff}$, and $2L_{\rm diff}$, with $L_{\rm Diff} = 4.99$ cm. We see that the shape of the dromion undergoes no apparent change, but its amplitude is reduced a little during propagation due to the small loss inherent in the dark oscillator q_2. With our system parameters, the average peak power of the plasmon dromion is $\bar{P}_{\rm peak} = 814$ mW, which corresponds average peak intensity $\bar{I}_{\rm peak} = 361$ mW/cm^2. We see that in the PIT metamaterial extremely low generation power is needed for generating (2+1)D spatiotemporal dromions.

6. Conclusions

We have proposed a scheme to realize giant Kerr nonlinearities and create low-power gigahertz solitons via PIT in a metamaterial. We have demonstrated that the PIT in such system can not only mimic the electromagnetically induced transparency in coherent three-level atomic systems, but also exhibit a crossover from PIT to Autler-Townes splitting. We have also shown that the system suggested here possess very large second-order and third-order nonlinear susceptibilities, which may abe used to create plasmonic solitons and dromions with extremely low generation power. Our studies raise the possibility for obtaining strong nonlinear effects for gigahertz radiations at very low intensity based on PIT metamaterials.

7. Acknowledgments

This work was supported by the NSF-China under Grants No. 11174080 and No. 11474099, and by the Chinese Education Ministry Reward for Excellent Doctors in Academics under Grant No. PY2014009.

References

[1] S. Zhang, D. A. Genov, Y. Wang, M. Liu, and X. Zhang, Phys. Rev. Lett. **101**, 047401 (2008)
[2] N. Papasimakis, V. A. Fedotov, N. I. Zheludev, and S. L. Prosvirnin, Phys. Rev. Lett. **101**, 253903 (2008)
[3] P. Tassin, L. Zhang, T. Koschny, E. N. Economou, and C. M. Soukoulis, Phys. Rev. Lett. **102**, 053901 (2009)
[4] R. Singh, C. Rockstuhl, F. Lederer, and W. Zhang Phys. Rev. B **79**, 085111 (2009)
[5] M. Fleischhauer, A. Imamoglu, and J. P. Marangos, Rev. Mod. Phys. **77**, 633 (2005)
[6] A. A. Zharov, I. V. Shadrivov, and Y. S. Kivshar, Phys. Rev. Lett. **91**, 037401 (2003)
[7] B. Wang, J. F. Zhou, T. Koschny, and C. M. Soukoulis, Opt. Express **16**, 16058-16063 (2008)
[8] Y. Sun, Y. Tong, C. Xue, Y. Ding, Y. Li, H. Jiang, and H. Chen, Appl. Phys. Lett. **103**, 091904 (2013)
[9] M. Lapine, I. V. Shadrivov, and Y. S. Kivshar, Rev. Mod. Phys. **86**, 1093 (2014)
[10] Z. Bai, C. Hang, and G. Huang, Opt. Express **21**, 17736 (2013)
[11] Z. Bai, G. Huang, L. Liu, and S. Zhang, Sci. Reports **5**, 13780 (2015)
[12] Z. Bai and G. Huang, Phys. Rev. A **93**, 013818 (2016)
[13] J. Gu, R. Singh, X. Liu, X. Zhang, Y. Ma, S. Zhang, S. A. Maier, Z. Tian, A. K. Azad, H.-T. Chen, A. J. Taylor, J. Han, and W. Zhang, Nat. Commun. **3**, 1151 (2012)
[14] C. Zhu, C. Tan, and G. Huang, Phys. Rev. A **87**, 043813 (2013)
[15] C. Tan, and G. Huang, J. Opt. Soc. Am. B **31**, 704 (2014)
[16] A. C. Newell and J. V. Moloney, *Nonlinear Optics* (Addison-Wesley, Redwood City, 1992), pages 74 and 75

Application of quantum control techniques to design broadband and ultra-broadband half-wave plates

Wei Huang and Elica Kyoseva

Engineering Product Development, Singapore University of Technology and Design, 8 Somapah Road, 487372 Singapore, Singapore

E-mail: elica_kyoseva@sutd.edu.sg

Abstract. Broadband and ultra-broadband half-wave plates with composite design were previously experimentally demonstrated by E. Dimova *et. al.*, in *Opt. Commun.* **366**, 382 (2016). To achieve broadband and ultra-broadband performance, the authors utilized the relative rotation between the individual identical half-wave plates as control parameters, thus simulating the well-known control technique of composite pulses from quantum optics. In this paper, we extend the broadband and ultra-broadband designs by relaxing the condition of having identical constituent half-wave plates and allow for their thicknesses to also be variable control parameters. We further relax the condition of having symmetric rotation angles and solve numerically for both control parameters - rotation angles and thicknesses of the birefringent plates. We show that the presented design is broadband over a much wider range of the wavelengths, compared to the original design, when using the same number of constituent half-wave plates.

1. Introduction

A half-wave plate is an optical device, which rotates the polarization direction of a linearly-polarized light at an angle 2θ, where θ denotes the angle between the incident light polarization direction and the fast axis of the half-wave plate [1, 2, 3]. A notable drawback of the original half-wave plate is that it works only for a narrow wavelength bandwidth of the incident light due to the strong wavelength dispersion. To solve this disadvantage, Clarke and Becker [4, 5] introduced an additional medium with reciprocal dispersion to reduce the wavelength dispersion. Another approach to reduce the dispersion is to use a stack of waveplates whose optical axes are rotated at given angles creating a sequence of phase shifts and thereby near constant retardation over a large bandwidth or for a larger set of discrete wavelengths [6, 7, 8].

Most recently, an analogy between the evolution of quantum systems and the space propagation of light allowed for the successful application of quantum control techniques to optics systems such as coupled waveguides [9, 10]. This analogy can also be extended to design broadband optical elements based on the broadband composite pulses technique widely used in NMR systems [11]. Thus far, several groups have proposed and experimentally demonstrated broadband polarization waveplates [12, 13, 14, 15, 16, 17, 18].

In this paper we reveal the applicability of the method of composite pulses to design modular broadband and ultra-broadband half-wave plates, irrespective of the optical medium used. We first give an overview of the existing designs of modular half-wave plates and show that they operate over a wide range of wavelengths and how their performance depends on the number of individual half-wave plates used. For the ultra-broadband half-wave plates, commercially available achromatic half-wave plates are utilized. Finally, we present a novel design of composite half-wave plates where the constituent birefringent plates have variable length. Using this additional variable as a control parameter, we design half-wave plates with significantly wider wavelength range of operation.

2. Theory

A single birefringent retarder rotated at an angle θ with respect to the slow and the fast axes of the retarder is described by the Jones matrix,

$$\mathfrak{J}_\theta(\varphi) = \mathfrak{R}(-\theta) \begin{bmatrix} e^{i\varphi/2} & 0 \\ 0 & e^{-i\varphi/2} \end{bmatrix} \mathfrak{R}(\theta), \tag{1}$$

where

$$\mathfrak{R}(\theta) = \begin{bmatrix} \cos\theta & \sin\theta \\ -\sin\theta & \cos\theta \end{bmatrix}. \tag{2}$$

Here, the phase retardance is $\varphi = 2\pi L(n_f - n_s)/\lambda$ with n_f and n_s being the refractive indices along the fast and slow axes of the birefringent plate, λ being the wavelength of the input light, and L - its length. When the phase retardance is $\varphi = \pi$, then the retarder is a standard half-wave plate.

We proceed to change the basis from the horizontal-vertical polarization basis (HV) to left-right circular polarization basis (LR) by using the transformation matrix $\mathbf{W}$, such that $\mathbf{J}_\theta(\varphi) = \mathbf{W}^{-1}\mathfrak{J}_\theta(\varphi)\mathbf{W}$, with

$$\mathbf{W} = \tfrac{1}{\sqrt{2}} \begin{bmatrix} 1 & 1 \\ -i & i \end{bmatrix}. \tag{3}$$

Thus, we obtain that the Jones matrix for a retarder with a phase retardance φ and rotated at an angle θ is given as (in the LR basis) [12, 13, 14, 15, 16, 17, 18],

$$\mathbf{J}_\theta(\varphi) = \begin{bmatrix} \cos(\varphi/2) & i\sin(\varphi/2)\, e^{2i\theta} \\ i\sin(\varphi/2)\, e^{-2i\theta} & \cos(\varphi/2) \end{bmatrix}. \tag{4}$$

Designing half-wave plates that are robust to variations in the phase retardance φ around values $\varphi = m\pi$, $(m \in \mathbb{N})$, is done in an analogy to composite pulses [13, 14, 15, 16], which is a widely-used technique for robust control in quantum optics [19, 20]. We replace the birefringent retarder (half-wave plate) with a stack of an odd number $N = (2n+1)$ of birefringent retarders, thus creating a half-wave plate with composite or modular design. Each individual birefringent retarder has a phase retardance $\varphi = \pi$. However, each is rotated at a different angle θ_k. To obtain the total Jones matrix in the LR basis of the above described stack of birefringent retarders (half-wave plates), we multiply the individual Jones matrices (4) according to (read from right to left),

$$\mathbf{J}^{(N)} = \mathbf{J}_{\theta_N}(\varphi)\, \mathbf{J}_{\theta_{N-1}}(\varphi) \cdots \mathbf{J}_{\theta_1}(\varphi). \tag{5}$$

We require that the total effect of the modular half-wave plate (5) is identical to one from an ideal half-wave plate propagator, which is given by the Jones matrix $\mathbf{J}_0$ in the LR basis (up to a global phase factor),

$$\mathbf{J}_0 = \begin{bmatrix} 0 & i \\ i & 0 \end{bmatrix}. \tag{6}$$

Mathematically, this corresponds to setting $\mathbf{J}^{(N)} \equiv \mathbf{J}_0$ at $\varphi = \pi$, which allows us to use the n independent angles θ_k as control parameters. After calculations [15, 18], we obtain that the following symmetry condition for the rotation angles $\theta_k = \theta_{N+1-k}$, $(k = 1, 2, ..., n)$ ensures

Table 1. For different number N of the individual half-wave plates, rotation angles θ_k (in degrees) assuring a broadband behaviour of the modular half-wave plate [15, 18].

N	Rotation angles (θ_1; θ_2; $\cdots$; θ_{N-1}; θ_N)
3	(60; 120; 60)
5	(51.0; 79.7; 147.3; 79.7; 51.0)
7	(68.0; 16.6; 98.4; 119.8; 98.4; 16.6; 68.0)
9	(99.4; 25.1; 64.7; 141.0; 93.8; 141.0; 64.7; 25.1; 99.4)

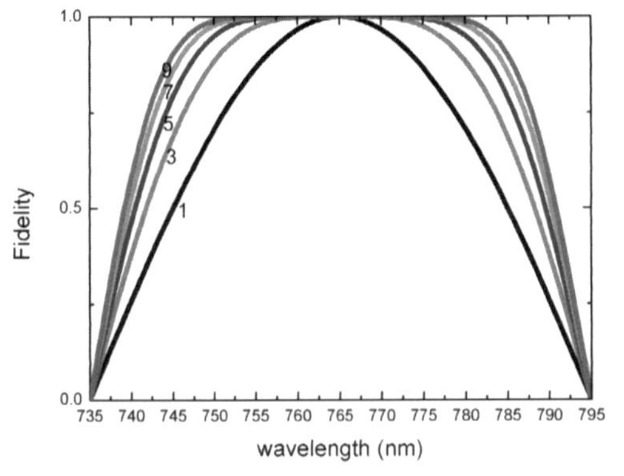

Figure 1. Calculated fidelity of the Jones matrix of the modular half-wave plate, $F = tr(\mathbf{J}_0^{-1}\mathbf{J}^{(N)})$, as a function of the wavelength λ. The number of individual half-wave plates (WPMQ10M-780, Thorlabs) is N={3, 5, 7, 9} and their respective angles of rotation are given in Table 1 [18]. The fidelity of a single half-wave plate is labeled by 1 and is added for easy comparison.

that the total effect of the modular half-wave plate is that of an ideal half-wave plate. Several examples of modular broadband half-wave plates arrangements are presented in Table 1.

As shown in Fig.1, the modular designs of half-wave plates have significantly extended spectral range of operation. To achieve an even wider range of operation, one should increase the number of individual half-wave plates used to assemble the modular half-wave plate. These results are proven in the experimental paper of Ref. [18] and all experimental details can be found there.

3. Modified broadband half-wave plate

We propose a new design of broadband and ultra-broadband half-wave plates, which is different in two significant ways. First, additionally to the rotation angles, we also assume to have the freedom to vary the length of the individual half-wave plates as control parameters, L_n. Thus, the individual half-wave plates have a variable phase retardance, φ by using $\varphi_n = 2\pi L_n(n_f - n_s)/\lambda$. Secondly, we relax the condition of having symmetry in the rotation angles, as shown in Table 1, and allow for arbitrary rotation angles. We assume that we use the same single half-wave plates (WPMQ10M-780, Thorlabs), as in Section 2. This scheme is given in Figure 2.

The Jones matrix $\bar{\mathbf{J}}^{(N)}$ of the above described arrangement of waveplates in the LR basis is given by (read from right to left):

$$\bar{\mathbf{J}}^{(N)} = \mathbf{J}_{\theta_N}(\varphi_N)\,\mathbf{J}_{\theta_{N-1}}(\varphi_{N-1})\cdots\mathbf{J}_{\theta_1}(\varphi_1)\,. \tag{7}$$

We proceed as in Section 2 and require that $\bar{\mathbf{J}}^{(N)} \equiv \mathbf{J}_0$. We seek to nullify as many lowest order derivatives of $\bar{\mathbf{J}}^{(N)}$ vs the optical retardance φ at $\varphi = \pi$ as possible. Thus, we can obtain the

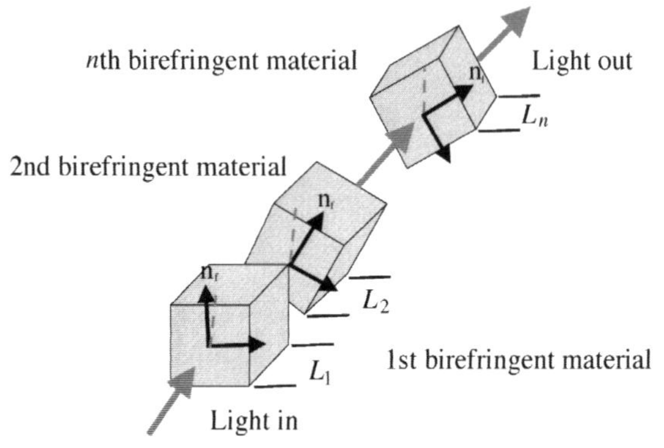

Figure 2. The schematics of the modified broadband half-wave plate with variable lengths L_n of the individual half-wave plates and different angles of rotation θ_n.

following nonlinear algebraic equations for the modified Jones matrix,

$$\left[\partial_{\varphi}^{k}\bar{\mathbf{J}}_{11}^{(N)}\right]_{\varphi=\pi}=0,\quad \left[\partial_{\varphi}^{k}\bar{\mathbf{J}}_{12}^{(N)}\right]_{\varphi=\pi}=0, \tag{8}$$

where $(k = 1, 2, ..., n)$.

We can solve these equations to obtain the rotation angles and the lengths of the individual half-wave plates. This method gives more freedom than the one described in Section 2. However, the trade off is that it is very difficult to find analytical solutions of Eqs. (8) and we proceed to solve them numerically. For simplicity, we express the lengths of the half-wave plates in terms of the phase retardance φ_n. Table 2 lists our results for the rotation angles θ_n and the phase retardance φ_n for $N = \{3, 5\}$.

Table 2. Rotation angles θ_k (in degrees) and phase retardance φ_k (in radians) of N individual half-wave plates to design broadband modular half-wave plates.

N	Rotation angles $(\theta_1;\ \cdots;\ \theta_N)$	Lengths of birefringence $(\varphi_1;\ \cdots;\ \varphi_N)$
3	(23.17; 148.70; 209.54)	$(0.735\pi;\ 0.682\pi;\ 0.693\pi)$
5	(233.56; 9.06; 316.09; 46.66; 204.29)	$(0.73\pi;\ 0.74\pi;\ 1.78\pi;\ 1.04\pi;\ 1.47\pi)$

We compare the performances of the broadband half-wave plates presented in Sections 2 and 3. We assume that the constituent half-wave plates have the same single half-wave plate fidelities, which have been extracted from experimental measurements (WPMQ10M-780, Thorlabs). In Fig. 3 we plot the results for $N = \{3, 5\}$ individual half-wave plates. It is clearly shown that the modified design of modular half-wave plate with variable length of the individual half-wave plates is broadband over a much wider range of the wavelength of the incident light compared to the previously described designs.

4. Conclusions

We demonstrated modular broadband half-wave plates and we proposed a novel design of even more broadband modular half-wave plates, which is achieved due to a large freedom in setting the variable control parameters. We calculated numerically the values of these parameters – rotation

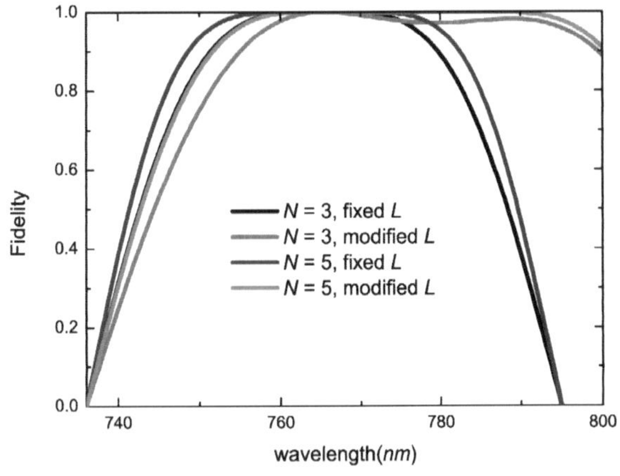

Figure 3. Fidelity $F = tr(\mathbf{J}_0^{-1}\bar{\mathbf{J}}^{(N)})$ of the Jones matrix of the modified broadband half-wave plate as a function of the wavelength λ. The number of individual half-wave plates is $N = \{3, 5\}$ and we compare with the results of the existing broadband half-wave plates from Fig. 1 also with $N = \{3,5\}$.

angles and lengths of the individual half-wave plates – for $N = \{3, 5\}$ number of constituent half-wave plates. We showed that by using the same number of individual half-wave plates we obtain a half-wave plates with a larger bandwidth compared to previous designs. Finally, our results confirm the applicability of the composite pulses control technique from quantum optics to design robust optical elements, irrespective of the birefringent materials used.

5. Acknowledgements
We acknowledge financial support by an SUTD-MIT International Design Centre Grant, Project No. IDG 31300102.

6. References
[1] Hecht E, 2002 *Optics* (Addison Wesley, San Francisco).
[2] Born M and Wolf E, 2005 *Principles of Optics* (Cambridge University Press, Cambridge).
[3] Azzam M A and Bashara N M, 1977 *Ellipsometry and Polarized Light* (North Holland, Amsterdam).
[4] Clarke D, 1967 Achromatic half-wave plates and linear polarization rotators *Opt. Acta* **14**, 343.
[5] Beckers J M, 1971 Achromatic Linear Retarders *Appl. Opt.* **10**, 973.
[6] Pancharatnam S, 1955 Achromatic combinations of birefringent plates *Proc. Indian Acad. Sci.* **41**, 130.
[7] Title A M, 1975 Improvement of Birefringent Filters. 2: Achromatic Waveplates *Appl. Opt.* **14**, 229.
[8] McIntyre C M and Harris S E, 1968 Achromatic Wave Plates for the Visible Spectrum *J. Opt. Soc. Am.* **58**, 1575.
[9] Longhi S, Della V G, Ornigotti M, and Laporta P, 2007 Coherent tunneling by adiabatic passage in an optical waveguide system *Phys. Rev. B.* **76** 201101.
[10] Huang W, Rangelov A A, and Kyoseva E, 2014 Complete achromatic optical switching between two waveguides with a sign flip of the phase mismatch *Phys. Rev. A.* **90** 053837.
[11] Levitt M H, 1986 Composite pulses *Prog. Nucl. Magn. Reson. Spectrosc.* **18** 61.
[12] Ardavan A, 2007 Exploiting the Poincare Bloch symmetry to design high-fidelity broadband composite linear retarders *New J. Phys.* **9**, 24.
[13] Ivanov S S, Rangelov A A, Vitanov N V, Peters T and Halfmann T, 2012 Highly efficient broadband conversion of light polarization by composite retarders *J. Opt. Soc. Am. A* **29**, 265.
[14] Peters T, Ivanov S S, Englisch D, Rangelov A A, Vitanov N V and Halfmann T, 2012 Variable ultrabroadband and narrowband composite polarization retarders *Appl. Opt.* **51**, 7466.
[15] Rangelov A A, 2012 Achromatic change of circular polarization handedness *Opt. Commun.* **285**, 4157.
[16] Rangelov A A and Kyoseva E, 2015 Broadband composite polarization rotator *Opt. Commun.* **338**, 574.

[17] Dimova E S, Ivanov S S, Popkirov G S and Vitanov N V, 2014 Highly efficient broadband polarization retarders and tunable polarization filters made of composite stacks of ordinary wave plates *J. Opt. Soc. Am. A* **31**, 952.

[18] Dimova E, Huang W, Popkirov G, Rangelov A and Kyoseva E, 2016 Broadband and ultra-broadband modular half-wave plates *Opt. Commun.* **366**, 382.

[19] Timoney N, Elman V, Glaser S, Weiss C, Johanning M, Neuhauser W and Wunderlich C, 2008 Error-resistant single-qubit gates with trapped ions *Phys. Rev. A* **77**, 052334.

[20] Häffner H, Roos C F and Blatt R, 2008 Quantum computing with trapped ions *Phys. Rep.* **469**, 155.

Basis entropy: A useful physical quantity about projective measurement

Xing Chen
Institute Of Physics, Chinese Academy Of Sciences, 100190 Beijing, People's Republic of China
E-mail: chenx@iphy.ac.cn

Abstract. Projective measurement can increase the entropy of a state ρ, the increased entropy is not only up to the basis of projective measurement, but also has something to do with the properties of the state itself. In this paper we define this increased entropy as basis entropy. And then we discuss the usefulness of this new concept by showing its application in deciding whether a state is pure or not and detecting the existence of quantum discord. And as shown in the paper, this new concept can also be used to describe decoherence.

1. Introduction

Projective measurement can increase the entropy of a state ρ[1][3]. And for different states, the increased entropy is different, and it is up to two factors

(i) The orthogonal projectors of projective measurement
(ii) The state itself.

Every state has its increased entropy after a projective measurement. This increased entropy is actually a quite useful physical quantity. The aim of this paper is to show the usefulness of this increased entropy by providing some theorems about this increased entropy. A mentionable merit of this increased entropy is that it is highly related to quantum discord, and it can be used to detect the existence of quantum discord.

2. Basis entropy

First we should give this increased entropy a proper name. Since projective measurement is dependent on its basis, we suggest using 'basis entropy' to describe this increased entropy of the state. The meaning of which is, with the knowledge of the state(the knowledge here means the information which equals to the von Neumann entropy of this state), the ignorance of a state's projective measurement result. And according to the definition of basis entropy, we can use the following formula to calculate basis entropy

$$BE = S(\sum_i P_i \rho P_i) - S(\rho) \tag{1}$$

where P_i is a complete set of orthogonal projectors.

A good example which can illustrate the physical meaning of basis entropy is Grove's algorithm[4]. First let us review the procedure of Grover's algorithm[1].

(i) $|0\rangle^{\otimes n}|1\rangle$

(ii) $\longrightarrow \frac{1}{\sqrt{2^n}}\sum_{x=0}^{2^n-1}|x\rangle[\frac{|0\rangle-|1\rangle}{\sqrt{2}}]$

(iii) $\longrightarrow [(2|\psi\rangle\langle\psi|-\mathbf{I})\mathbf{O}]^R\frac{1}{\sqrt{2^n}}\sum_{x=0}^{2^n-1}|x\rangle[\frac{|0\rangle-|1\rangle}{\sqrt{2}}]$
$\approx |x_0\rangle[\frac{|0\rangle-|1\rangle}{\sqrt{2}}]$

(iv) $\longrightarrow x_0$

The first state is the initial state, where n represents the qubit number, by applying $H^{\otimes n}$ to initial n qubits we get the state in step two, and then we apply the Grover iteration $(2|\psi\rangle\langle\psi|-\mathbf{I})\mathbf{O}$[6], we get the state in step three. After applying the Grover iteration for about $R \approx \lceil \pi\sqrt{2^n}/4 \rceil$ times, we get the wanted state $|x_0\rangle$, the detail of Grover's algorithm refer [1].

For Grover's algorithm, we actually search in the state $(\sum_{i=0}^{2^n-1}|x_i\rangle)/\sqrt{2^n}$, and measure it with projectors $\{|0\rangle\langle 0|, |1\rangle\langle 1|\}$, before apply Grover iteration, the basis entropy of the database state is n, this is our ignorance of the measurement result, which means without applying Grover iteration, we need n bits to describe the measurement result. After apply one time of Grover iteration, the basis entropy decreased, which means our ignorance of the measurement result decreased, so the probability of finding the target state $|x_0\rangle$ increased. As shown in Figure 1, with the decreasing of the basis entropy, the successful probability is increasing.

Figure 1 is the basis entropy change of Grover's algorithm, here we set $n = 20$, clearly we can see from this figure that, the more time we apply Grover iteration(under the desired time $R \approx \lceil \pi\sqrt{2^n}/4 \rceil$, see appendix A), the basis entropy will be smaller and the success probability will be higher.

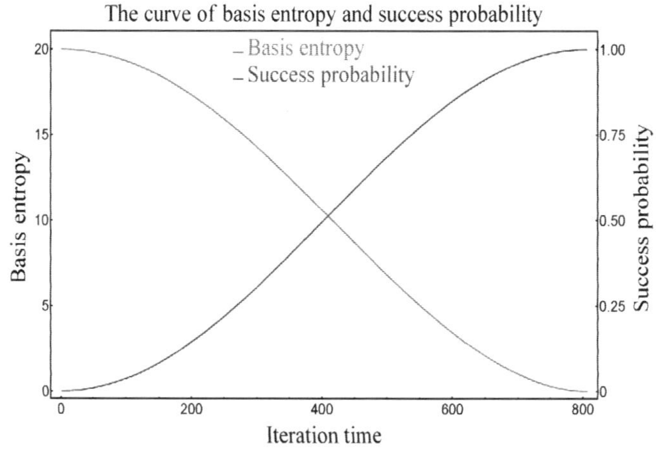

Figure 1: *Basis entropy and success probability change of Grover's algorithm*

Another example is Shor's algorithm. The procedure of Shor's algorithm is shown as follows [1]:

(i) $|0\rangle|1\rangle$ initial state

(ii) $\longrightarrow \frac{1}{\sqrt{2^t}}\sum_{j=0}^{2^t-1}|j\rangle|1\rangle$ create superposition

(iii) $\longrightarrow \frac{1}{\sqrt{2^t}} \sum_{j=0}^{2^t-1} \lvert j\rangle \lvert x^j modN\rangle$ $\approx \frac{1}{r2^t} \sum_{s=0}^{r-1} \sum_{j=0}^{2^t-1} e^{2\pi i s j/r} \lvert j\rangle \lvert u_s\rangle$	apply $U_{x,N}$
(iv) $\longrightarrow \frac{1}{\sqrt{r}} \sum_{s=0}^{r-1} \lvert \widetilde{s/r}\rangle \lvert u_s\rangle$	apply inverse Fourier transform to first register
(v) $\longrightarrow \widetilde{s/r}$	measure first register
(vi) $\longrightarrow r$	apply continued fractions algorithm

where $U_{x,N}$ performs the transformation $\lvert j\rangle\lvert k\rangle \longrightarrow \lvert j\rangle\lvert x^j k\ modN\rangle$. r is the least integer such that $x^r = 1(modN)$, Figure 2 is the quantum circuit of Shor's algorithm.

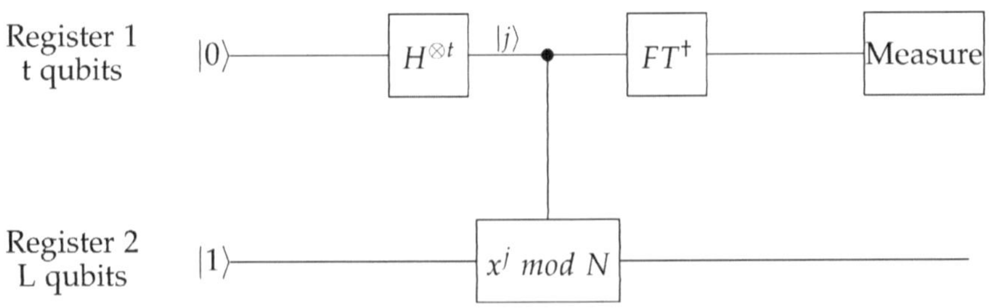

Figure 2: *Quantum circuit of Shor's algorithm*

where t is the qubit number of $\lvert 0\rangle$, and L is the qubit number of $\lvert 1\rangle$. For more details of Shor's algorithm, refer [1]. Next, we will analyse the basis entropy of Shor's algorithm.

In Shor's algorithm, we use state $(\sum_{j=0}^{2^t-1} \lvert j\rangle\lvert 1\rangle)/\sqrt{2^t}$ to factor great numbers, but we get the useful measurement result from the first register, so we only calculate the basis entropy of the first register. The measurement projectors of Shor's algorithm are $\{\lvert 0\rangle\langle 0\rvert, \lvert 1\rangle\langle 1\rvert\}$. Before we apply Shor's algorithm, the basis entropy of the first register is t, and then after the $U_{x,N}$ transformation, the basis entropy remains unchanged, and after the inverse Fourier transform, the basis entropy becomes r, then the main part of Shor's algorithm is over, the subsequent steps are measuring the state and applying mathematical methods to the measurement result.

From the change of the basis entropy we can easily see that the most crucial step of Shor's algorithm is applying inverse Fourier transform, because this step reduced the basis entropy of the first register without decohering it. After this crucial step, the basis entropy of the first register reduced from t to r, and then we have a higher probability to get the desired state by measuring the first register.

From the two examples above we can see that basis entropy is the ignorance of the measurement result with the knowledge of the state.

3. Maximum and minimum basis entropy

A state's basis entropy is dependent on different measurement projectors, so the number of its basis entropy is actually infinite, but we are only interested in its maximum and minimum basis entropy.

3.1. The property of maximum basis entropy

For a state's maximum basis entropy, we have the following theorem.

Theorem 1. Iff a state's maximal basis entropy is $log_2 D$, the state is a pure state.

where D is the dimension of Hilbert space.

Proof: If a state's maximal basis entropy is $log_2 D$, which means

$$BE_{max} = S(\sum_i P_i \rho P_i) - S(\rho) = log_2 D \tag{2}$$

Since the entropy of a state is non-negative and in a D-dimensional Hilbert space, the state's von Neumann entropy is at most $log_2 D$[1], so if the basis entropy of a state is $log_2 D$, then $S(\rho)$ must be zero, which means the state is a pure state.

Let's prove the other direction of the theorem. It seems only for pure state like

$$|\psi\rangle = \frac{1}{\sqrt{D}} \sum_i^D |i\rangle \tag{3}$$

has $log_2 D$ basis entropy, while for pure state like $(\sqrt{3}|0\rangle + |1\rangle)/2$, its basis entropy is smaller than $log_2 D$. We will prove that for any pure state, its maximum basis entropy will be $log_2 D$, as long as we choose right projectors to measure them. For simplicity, we just prove the case $D = 2$. For any pure state, we can write its density matrix as

$$\rho = \frac{1}{2}\mathbf{I} + a\sigma_1 + b\sigma_2 + c\sigma_3 \tag{4}$$

where a, b and c are coefficients and $\sigma_1, \sigma_2, \sigma_3$ are Pauli matrices. We only need to prove that

$$S_{max}(\sum_i^D P_i \rho P_i) = 1 \tag{5}$$

where,

$$\{P_k = V|k\rangle\langle k|V^{\dagger} : k = 0, 1\} \tag{6}$$

is the complete set of orthogonal projectors. And V is a 2-dimensional unitary transformation. It has be proven that for any 2-dimensional pure state, its maximum basis entropy is $log_2 2 = 1$ (for those who are interested in the details ,see appendix B). And now we complete the proof of theorem 1. The proof can be easily extended to higher Hilbert space.

From theorem 1, it is natural to get the following corollary:

Corollary 1. Only for states like $\rho = \sum_i^D |i\rangle\langle i|/D$, they have no basis entropy, it means no projective measurement can increase their entropy.

This corollary is easy to prove, since for a state like $\rho = \sum_i^D |i\rangle\langle i|/D$, its von Neumann has reached the maximal value in its Hilbert space, so no projective measurement can increase its entropy.

We can use maximum basis entropy to judge whether a state is mixed or not. According to theorem 1, if a state's maximum basis entropy is $log_2 D$, then the state is a pure state. If its maximum basis entropy is smaller than $log_2 D$, then it's a mixed state. If its maximum basis entropy is zero, it has no basis entropy, then the state is $\rho = \sum_i^D |i\rangle\langle i|/D$.

3.2. *The property of minimum basis entropy*

For most states, regardless of pure states or mixed states, their minimum basis entropy is zero, which means there exists a complete set of orthogonal projectors, under these projectors, we can get the full knowledge of this state. But there are some states, their minimum basis entropy are not zero, which means no matter what projectors we use to measure them, there is always some information we can not get. This inaccessible information, we will show next, is related to quantum discord.

Quantum discord comes from the projective measurement[7], just as the basis entropy. For a bell state, its basis entropy is a constant, so it's easy for us to calculate its minimum basis entropy. So let's take a bell state as an example to explain why the quantum discord is related to minimum basis entropy. For bell state like $(|00\rangle + |11\rangle)/\sqrt{2}$, the subsystem is A and B, the discord of this state is

$$\begin{aligned}\delta(A:B)_{\{\Pi_i^B\}} =& I(A:B) - J(A:B)_{\{\Pi_i^B\}}\\ =& S(A) + S(B) - S(A,B) - S(A) + S(A|\Pi_i^B)\\ =& S(B) + S(A|\Pi_i^B)\\ =& S(B)\\ =& 1\end{aligned} \tag{7}$$

where $I(A:B)$ is the quantum mutual information between A and B, and $J(A:B)_{\{\Pi_i^B\}}$ is the measurement mutual information between A and B. From this equation we can see that the quantum discord of this state comes from $S(B)$. Then let's calculate the basis entropy of the joint system AB, measure AB by projectors $\{|00\rangle\langle 00|, |01\rangle\langle 01|, |10\rangle\langle 10|, |11\rangle\langle 11|\}$, and after this projective measurement the whole system AB will become

$$\rho_{AB_{pm}} = \frac{1}{2}\begin{pmatrix}1&0&0&0\\0&0&0&0\\0&0&0&0\\0&0&0&1\end{pmatrix} \tag{8}$$

and

$$BE_{\beta_{00}} = S(\rho_{AB_{pm}}) - S(\rho_{AB}) = 1 = S(B) \tag{9}$$

Therefore the basis entropy of the joint system AB is equal to the quantum discord between its subsystems.

For another example, let's consider the discord of Werner state[8] $\rho = (1-z)\mathbf{I}/4 + z|\psi\rangle\langle\psi|$ with $|\psi\rangle = (|00\rangle + |11\rangle)/\sqrt{2}$. One can calculate that the quantum discord of this Werner state is equal to its basis entropy. For example as shown in Figure 3, when $z = 1/3$, the value of quantum discord is 0.1258, it's basis entropy is also 0.1258.

The quantum discord of Bell states or Werner state are exactly equal to their basis entropy, no matter which orthogonal projectors we used to measure them, like $\{|0\rangle\langle 0|, |1\rangle\langle 1|\}$ or $\{|+\rangle\langle +|, |-\rangle\langle -|\}$. This equivalence has two reasons. First, Bell states or Werner state are invariant under local rotations, so their basis entropy is invariant under different orthogonal projectors; second, for bell state or werner states, they are states with the maximally mixed marginals, thus their quantum discord is equal to their minimum basis entropy. This is also true for ordinary two-qubit system $\rho = (\mathbf{I} + \sum_{i=1}^{3} c_i\sigma_i \otimes \sigma_i)/4$[9] with maximum mixed marginals, its basis entropy is not a constant. We have shown in appendix C that, its quantum discord is still equal to its minimum basis entropy. For states without maximum marginals, its nonzero minimum basis entropy can guarantee its nonzero quantum discord, so we have the following theorem

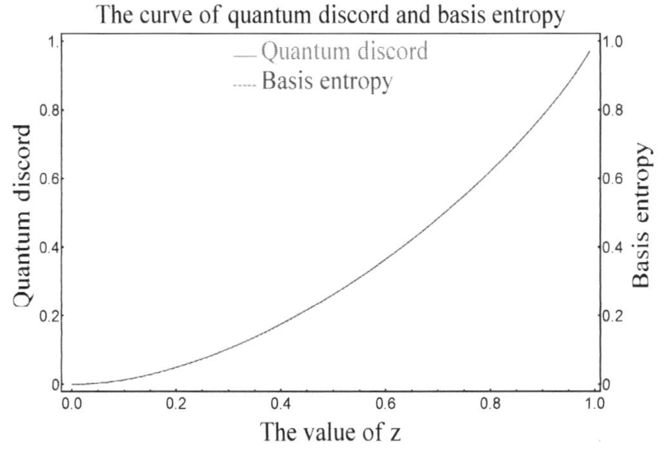

Figure 3: *Quantum discord and basis entropy of Werner state*

Theorem 2. If a state's minimum basis entropy is nonzero, there must be quantum discord between its subsystems.

From the above example we can see that, quantum discord is a symmetric physical quantity for states like $\rho = (\mathbf{I} + \sum_{i=1}^{3} c_i \sigma_i \otimes \sigma_i)/4$. For asymmetric quantum discord, the theorem still holds, see discussion in appendix C

4. Basis entropy and coherence

One can see that basis entropy is also highly related to quantum coherence[10]. Except the formula we use to measure basis entropy showed up naturally. One can find that the formula use to quantify coherence is actually a special case of basis entropy, it's a basis entropy with fixed basis, which means under a specific basis, there is no difference between basis entropy and coherence. But we can easily see that if a state has no coherence, we can't guarantee its maximum basis entropy is zero; if a state's maximum basis entropy is zero, then it must have no coherence. This difference between these two concepts has physical meaning. For example for state $|0\rangle$, it has no coherence, we can't decohere it, since its density matrix has no off-diagonal elements. But this state's maximum basis entropy is 1, from this perspective, we can still further decohere it.

The decoherence about coherence is basis-dependent[14], which means the state can only be seen as decohered under specific basis. The decoherence about maximum basis entropy is not basis-dependent, which means if we decohere a state's maximum basis entropy, the state will be fully decohered and it can't be further decohered. Take the following state as an example.

$$\rho_0 = \begin{pmatrix} 3/4 & \sqrt{3}/2 \\ \sqrt{3}/2 & 1/4 \end{pmatrix} \tag{10}$$

We decohere this state under basis $\{|0\rangle, |1\rangle\}$, it will become

$$\rho_0' = \begin{pmatrix} 3/4 & 0 \\ 0 & 1/4 \end{pmatrix} \tag{11}$$

According to corollary 1, this state has basis entropy, so it can be further decohered. The corresponding projectors are

$$P_1 = \frac{1}{2}\begin{pmatrix} 1 & -i \\ i & 1 \end{pmatrix}, P_2 = \frac{1}{2}\begin{pmatrix} 1 & i \\ -i & 1 \end{pmatrix} \tag{12}$$

After this , the state becomes

$$\rho_0'' = \begin{pmatrix} 1/2 & 0 \\ 0 & 1/2 \end{pmatrix} \tag{13}$$

According to corollary 1, this state has no basis entropy, so it can't be further decohered.

Therefore the maximum basis entropy can be used as a criterion to measure how much the environment can maximum decohere a quantum system, for pure states, they are highly isolated states and they have the maximum basis entropy, so they are easy to be decohered. And for state like $\rho = \sum_i^D |i\rangle\langle i|/D$, is actually a classical state, its maximum basis entropy is zero, so it has been fully decohered, the environment can't further decohere it.

5. Conclusions

In this paper we have shown that, the increased entropy of a state after a projective measurement is a useful physical quantity, we named this increased entropy as basis entropy and then we showed some usefulness of this new concept. The maximum basis entropy can be used to decide whether a state is pure or not, and the minimum basis entropy can be used to detect the existence of quantum discord. At the end of this paper, we also discussed the difference between basis entropy and coherence, and showed that basis entropy can be used to describe decoherence.

Acknowledgments

This research was supported by NSF of China, Grant No.11475254. And the help from Qikai He was gratefully acknowledged.

Appendix A

The success probability of Grover's algorithm: The initial search state is $\frac{1}{\sqrt{2^n}}\sum_{x=0}^{2^n-1}|x\rangle$, the target state is $|x_0\rangle$, we can rewrite the initial state as the combination of target state and un-target states. First of all let's define two normalized states

$$\begin{aligned} |\alpha\rangle &\equiv \frac{1}{\sqrt{2^n-1}}\sum_{\substack{x=0 \\ x\neq x_0}}^{2^n-1}|x\rangle \\ |\beta\rangle &\equiv |x_0\rangle \end{aligned} \tag{A.1}$$

where $|\alpha\rangle$ represents a sum over all x which are not the state we're searching for, $|\beta\rangle$ represents the target state $|x_0\rangle$. Then the initial search state can be write

$$\begin{aligned} |\psi\rangle &= \frac{1}{\sqrt{2^n}}\sum_{x=0}^{2^n-1}|x\rangle \\ &= \sqrt{\frac{2^n-1}{2^n}}|\alpha\rangle + \frac{1}{\sqrt{2^n}}|\beta\rangle \end{aligned} \tag{A.2}$$

As mentioned in [1], we can define $cos(\theta/2) = \sqrt{(2^n - 1)/2^n}$, then $|\psi\rangle$ can be write as $|\psi\rangle = cos(\theta/2)|\alpha\rangle + sin(\theta/2)|\beta\rangle$. Then after applying the Grover iteration $G = (2|\psi\rangle\langle\psi| - \mathbf{I})\mathbf{O}$[1] the state becomes

$$G|\psi\rangle = cos\frac{3\theta}{2}|\alpha\rangle + sin\frac{3\theta}{2}|\beta\rangle. \tag{A.3}$$

After applying G for k times, the state becomes

$$G^k|\psi\rangle = cos\frac{2k+1}{2}\theta|\alpha\rangle + sin\frac{2k+1}{2}\theta|\beta\rangle. \tag{A.4}$$

Then the success probability is the square of the amplitude of $|\beta\rangle$

$$p_{success} = sin^2[\frac{2k+1}{2}\theta] \tag{A.5}$$

The probability $p_{success}$ reaches its maximum when $k = k_{max}$, k_{max} should satisfy the following condition:

$$(2k_{max} + 1)\frac{\theta}{2} = \frac{\pi}{2} \tag{A.6}$$

considering k must be an integer, so we get $k_{max} = \lceil \frac{\pi}{2\theta} - \frac{1}{2} \rceil$. Generally, 2^n is very large and thus θ is very small, so the following approximation is acceptable

$$\theta \approx sin\theta \approx= \frac{2}{\sqrt{2^n}} \tag{A.7}$$

so $k_{max} \approx \lceil \pi\sqrt{2^n}/4 \rceil$, this is the desired time we mentioned in the paper before. It is obvious that under the desired time, with the increase of k, the success probability $p_{success}$ is increasing. For $n = 20$, the desired time is approximately 805, as shown in Figure 1.

The basis entropy of Grover's algorithm: Since we're searching in the state $|\psi\rangle = (\sum_{i=0}^{2^n-1} |x_i\rangle)/\sqrt{2^n}$, so we begin with this state. It's not difficult to check that under measurement projectors $\{|0\rangle\langle 0|, |1\rangle\langle 1|\}$, its basis entropy is n. No matter how many times we apply the Grover iteration, the state $G^k|\psi\rangle$ is a pure state, its von Neumann entropy is zero, so we just need to calculate the entropy of the state after projective measurement, which is

$$\rho_{pm} = |0\rangle\langle 0|G^k|\psi\rangle\langle\psi|G^k|0\rangle\langle 0| + |1\rangle\langle 1|G^k|\psi\rangle\langle\psi|G^k|1\rangle\langle 1| \tag{A.8}$$

which can be write as

$$\rho_{pm} = \begin{pmatrix} cos^2\frac{2k+1}{2}/(2^n-1) & & & & & \\ & \ddots & & & & \\ & & cos^2\frac{2k+1}{2}/(2^n-1) & & & \\ & & & sin^2\frac{2k+1}{2}/2^n & & \\ & & & & cos^2\frac{2k+1}{2}/(2^n-1) & \\ & & & & & \ddots \\ & & & & & & cos^2\frac{2k+1}{2}/(2^n-1) \end{pmatrix}_{2^n \times 2^n}$$

so we get

$$S(\rho_{pm}) = -[cos^2\frac{2k+1}{2}log_2(cos^2\frac{2k+1}{2}/(2^n-1)) + sin^2\frac{2k+1}{2}/2^n log_2(sin^2\frac{2k+1}{2}/2^n)] \tag{A.9}$$

Thus

$$\begin{aligned} BE_{grover} =& S(\rho_{pm}) - S(G^k|\psi\rangle\langle\psi|G^k) \\ =& -[cos^2\frac{2k+1}{2}log_2(cos^2\frac{2k+1}{2}/(2^n-1)) + sin^2\frac{2k+1}{2}/2^n log_2(sin^2\frac{2k+1}{2}/2^n)] \end{aligned} \tag{A.10}$$

With the increase of k, the basis entropy of the state $G^k|\psi\rangle$ is decreasing, while the success probability is increasing, as shown in Figure 1.

Appendix B

In this appendix, we will prove that for any 2-dimensional pure state, its maximum basis entropy is 1. For any 2-dimensional pure state, we can write its density matrix as

$$\rho = \frac{1}{2}\mathbf{I} + a\sigma_1 + b\sigma_2 + c\sigma_3 \tag{B.1}$$

where a, b, c are coefficients and $\sigma_1, \sigma_2, \sigma_3$ are pauli matrices. The measurement projectors are

$$\{B_k = V\Pi_k V^\dagger : k = 0, 1\} \tag{B.2}$$

where $\{\Pi_k = |k\rangle\langle k| : k = 0, 1\}$, $V \in U(2)$. And V can be written as $V = t\mathbf{I} + i\overrightarrow{y}\overrightarrow{\sigma}$, with $t \in R$, $\overrightarrow{y} = (y_1, y_2, y_3) \in R^3$, and $t^2 + y_1^2 + y_2^2 + y_3^2 = 1$. We only need to prove that

$$S(B_1\rho B_1 + B_2\rho B_2) = 1. \tag{B.3}$$

After calculation we get

$$\rho_{pm} = B_1\rho B_1 + B_2\rho B_2 = \frac{1}{2}\mathbf{I} + (az_1 + bz_2 + cz_3)(z_1\sigma_1 + z_2\sigma_2 + z_3\sigma_3) \tag{B.4}$$

where

$$\begin{aligned} z_1 &= 2(-ty_2 + y_1y_3); \\ z_2 &= 2(ty_1 + y_2y_3); \\ z_3 &= t^2 + y_3^2 - y_1^2 - y_2^2. \end{aligned} \tag{B.5}$$

After further calculation, we get the eigenvalues of the ρ_{pm}

$$\begin{aligned} \lambda_1 = \frac{1}{2} - &\sqrt{(}a^2z_1^4 + a^2z_2^2z_1^2 + a^2z_3^2z_1^2 + 2abz_2z_1^3 + 2abz_2^3z_1 + 2abz_2z_3^2z_1 + 2acz_3z_1^3 + 2acz_3^3z_1 + 2acz_2^2z_3z_1 \\ &+ b^2z_2^2z_1^2 + b^2z_2^4 + b^2z_2^2z_3^2 + 2bcz_2z_3z_1^2 + 2bcz_2z_3^3 + 2bcz_2^3z_3 + c^2z_3^2z_1^2 + c^2z_3^4 + c^2z_2^2z_3^2); \\ \lambda_2 = \frac{1}{2} + &\sqrt{(}a^2z_1^4 + a^2z_2^2z_1^2 + a^2z_3^2z_1^2 + 2abz_2z_1^3 + 2abz_2^3z_1 + 2abz_2z_3^2z_1 + 2acz_3z_1^3 + 2acz_3^3z_1 + 2acz_2^2z_3z_1 \\ &+ b^2z_2^2z_1^2 + b^2z_2^4 + b^2z_2^2z_3^2 + 2bcz_2z_3z_1^2 + 2bcz_2z_3^3 + 2bcz_2^3z_3 + c^2z_3^2z_1^2 + c^2z_3^4 + c^2z_2^2z_3^2) \end{aligned} \tag{B.6}$$

In order to make $S(\rho_{pm}) = 1$, we need to make sure

$$\begin{aligned} &\sqrt{(}a^2z_1^4 + a^2z_2^2z_1^2 + a^2z_3^2z_1^2 + 2abz_2z_1^3 + 2abz_2^3z_1 + 2abz_2z_3^2z_1 + 2acz_3z_1^3 + 2acz_3^3z_1 + 2acz_2^2z_3z_1 \\ &+ b^2z_2^2z_1^2 + b^2z_2^4 + b^2z_2^2z_3^2 + 2bcz_2z_3z_1^2 + 2bcz_2z_3^3 + 2bcz_2^3z_3 + c^2z_3^2z_1^2 + c^2z_3^4 + c^2z_2^2z_3^2) = 0. \end{aligned} \tag{B.7}$$

Using (B.5) and the following condition

$$t^2 + y_1^2 + y_2^2 + y_3^2 = 1 \tag{B.8}$$

we get formula (B.7) equals $2a(-ty_2 + y_1y_3) + 2b(ty_1 + y_2y_3) + c(t^2 + y_3^2 - y_1^2 - y_2^2)$, then we only need to make sure $2a(-ty_2 + y_1y_3) + 2b(ty_1 + y_2y_3) + c(t^2 + y_3^2 - y_1^2 - y_2^2) = 0$. Clearly, this equation has qualified solutions, for example, one of the solutions is

$$\begin{aligned}
t = {} & \frac{1}{ac}\left(2a^2\sqrt{\frac{a^2}{P} + \frac{c^2}{P} - \frac{bc}{P} - \frac{\sqrt{a^4 - 2a^2bc}}{P}} - bc\sqrt{\frac{a^2}{P} + \frac{c^2}{P} - \frac{bc}{P} - \frac{\sqrt{a^4 - 2a^2bc}}{P}} + \right. \\
& c^2\sqrt{\frac{a^2}{P} + \frac{c^2}{P} - \frac{bc}{P} - \frac{\sqrt{a^4 - 2a^2bc}}{P}} - 4a^2\left(\frac{a^2}{P} + \frac{c^2}{P} - \frac{bc}{P} - \frac{\sqrt{a^4 - 2a^2bc}}{P}\right)^{3/2} - \\
& 2b^2\left(\frac{a^2}{P} + \frac{c^2}{P} - \frac{bc}{P} - \frac{\sqrt{a^4 - 2a^2bc}}{P}\right)^{3/2} + 4bc\left(\frac{a^2}{P} + \frac{c^2}{P} - \frac{bc}{P} - \frac{\sqrt{a^4 - 2a^2bc}}{P}\right)^{3/2} - \\
& \left. 2c^2\left(\frac{a^2}{P} + \frac{c^2}{P} - \frac{bc}{P} - \frac{\sqrt{a^4 - 2a^2bc}}{P}\right)^{3/2}\right); \\
y_1 = {} & 0; \\
y_2 = {} & \sqrt{\frac{a^2}{P} + \frac{c^2}{P} - \frac{bc}{P} - \frac{\sqrt{a^4 - 2a^2bc}}{P}}; \\
y_3 = {} & \sqrt{\frac{a^2}{P} + \frac{c^2}{P} - \frac{bc}{P} - \frac{\sqrt{a^4 - 2a^2bc}}{P}}.[
\end{aligned} \tag{B.9}$$

where $P = 2(2a^2 + b^2 - 2bc + c^2)$ And then we can get

$$V = \begin{pmatrix} t + iy_3 & iy_1 + y_2 \\ iy_1 - y_2 & t - iy_3 \end{pmatrix} \tag{B.10}$$

Then measure $\rho = \frac{1}{2}\mathbf{I} + a\sigma_1 + b\sigma_2 + c\sigma_3$ by projectors $\{B_k = V\Pi_k V^\dagger : k = 0, 1\}$, we can get $S(B_1\rho B_1 + B_2\rho B_2) = 1$

Appendix C

For an ordinary two-qubit state like

$$\begin{aligned}
\rho = {} & \frac{1}{4}(\mathbf{I} + \sum_{i=1}^{3} c_i\sigma_i \otimes \sigma_i) \\
= {} & \frac{1}{4}\begin{pmatrix} 1 + c_3 & 0 & 0 & c_1 - c_2 \\ 0 & 1 - c_3 & c_1 + c_2 & 0 \\ 0 & c_1 + c_2 & 1 - c_3 & 0 \\ c_1 - c_2 & 0 & 0 & 1 + c_3 \end{pmatrix}
\end{aligned} \tag{C.1}$$

As shown in [9], the quantum discord of this state is

$$\begin{aligned} Q =& I(\rho) - C(\rho) \\ =& \frac{1}{4}[(1-c_1-c_2-c_3)log_2(1-c_1-c_2-c_3) \\ &+(1-c_1+c_2+c_3)log_2(1-c_1+c_2+c_3) \\ &+(1+c_1-c_2+c_3)log_2(1+c_1-c_2+c_3) \\ &+(1+c_1+c_2-c_3)log_2(1+c_1+c_2-c_3)] \\ &-\frac{1-c}{2}log_2(1-c)-\frac{1+c}{2}log_2(1+c) \end{aligned} \tag{C.2}$$

where c is defined as $c := max\{|c_1|,|c_2|,|c_3|\}$, $I(\rho)$ is quantum mutual information, and $C(\rho)$ is measurement mutual information. The basis entropy of this state is

$$BE = S(\sum_j B_j\rho B_j) - S(\rho) \tag{C.3}$$

where $\{B_j = V\Pi_k V^{\dagger} \otimes V\Pi_l V^{\dagger} : j = 0,1,2,3; k = 0,1; l = 0,1\}$, $\{\Pi_k = |k\rangle\langle k| : k = 0,1\}$, $V \in U(2)$. And V can be written as $V = t\mathbf{I} + i\overrightarrow{y}\overrightarrow{\sigma}$, with $t \in R$, $\overrightarrow{y} = (y_1,y_2,y_3) \in R^3$, and $t^2+y_1^2+y_2^2+y_3^2 = 1$. Which means

$$\begin{aligned} \rho_{pm} =& \sum_j B_j\rho B_j \\ =& \frac{1}{4}(\mathbf{I} + \sum_i\sum_k\sum_l c_i V\Pi_k V^{\dagger}\sigma_i V\Pi_k V^{\dagger} \otimes V\Pi_l V^{\dagger}\sigma_i V\Pi_l V^{\dagger}) \end{aligned} \tag{C.4}$$

After calculation, we get

$$\rho_{pm} = \frac{1}{4}(\mathbf{I} + (c_1z_1^2 + c_2z_2^2 + c_3z_3^2)V\sigma_3V^{\dagger} \otimes V\sigma_3V^{\dagger}) \tag{C.5}$$

where $z_1 = 2(-ty_2+y_1y_3), z_2 = 2(ty_1+y_2y_3), z_3 = t^2+y_3^2-y_1^2-y_2^2$. The eigenvalues of this ρ_{pm} is

$$\begin{aligned} \mu_1 =& \frac{1-c_1z_1^2-c_2z_2^2-c_3z_3^2}{4} \\ \mu_2 =& \frac{1-c_1z_1^2-c_2z_2^2-c_3z_3^2}{4} \\ \mu_3 =& \frac{1+c_1z_1^2+c_2z_2^2+c_3z_3^2}{4} \\ \mu_4 =& \frac{1+c_1z_1^2+c_2z_2^2+c_3z_3^2}{4} \end{aligned} \tag{C.6}$$

while the eigenvalues of ρ is

$$\begin{aligned} \lambda_1 =& \frac{1-c_1-c_2-c_3}{4} \\ \lambda_2 =& \frac{1-c_1+c_2+c_3}{4} \\ \lambda_3 =& \frac{1+c_1-c_2+c_3}{4} \\ \lambda_4 =& \frac{1+c_1+c_2-c_3}{4} \end{aligned} \tag{C.7}$$

So the basis entropy will be

$$\begin{aligned} S(\rho_{pm}) - S(\rho) =& \frac{1}{4}[(1-c_1-c_2-c_3)log_2(1-c_1-c_2-c_3) \\ &+(1-c_1+c_2+c_3)log_2(1-c_1+c_2+c_3) \\ &+(1+c_1-c_2+c_3)log_2(1+c_1-c_2+c_3) \\ &+(1+c_1+c_2-c_3)log_2(1+c_1+c_2-c_3)] \\ &-\frac{1-c_1z_1^2-c_2z_2^2-c_3z_3^2}{2}log_2(1-c) - \frac{1+c_1z_1^2+c_2z_2^2+c_3z_3^2}{2}log_2(1+c) \end{aligned} \tag{C.8}$$

Make $c = max\{|c_1|, |c_2|, |c_3|\}$ and notice that $z_1^2+z_2^2+z_3^2 = 1$, we get the minimum basis entropy:

$$\begin{aligned} S(\rho_{pm}) - S(\rho) =& \frac{1}{4}[(1-c_1-c_2-c_3)log_2(1-c_1-c_2-c_3) \\ &+(1-c_1+c_2+c_3)log_2(1-c_1+c_2+c_3) \\ &+(1+c_1-c_2+c_3)log_2(1+c_1-c_2+c_3) \\ &+(1+c_1+c_2-c_3)log_2(1+c_1+c_2-c_3)] \\ &-\frac{1-c}{2}log_2(1-c) - \frac{1+c}{2}log_2(1+c) \end{aligned} \tag{C.9}$$

which is exactly the quantum discord between subsystems A and B.

Quantum discord between subsystems in state $\rho = (\mathbf{I} + \sum_{i=1}^{3} c_i\sigma_i \otimes \sigma_i)/4$ is a symmetric quantity, but for others joint states, quantum discord may not be a symmetric quantity. For example, for state like

$$\rho = \frac{1}{4}\begin{pmatrix} 2 & 0 & 0 & 0 \\ 0 & 0 & 0 & 0 \\ 0 & 0 & 1 & 1 \\ 0 & 0 & 1 & 1 \end{pmatrix} \tag{C.10}$$

The quantum discord $\delta(A:B)_{\{\Pi_i^B\}}$ is nonzero, while $\delta(B:A)_{\{\Pi_i^A\}}$ is zero. Although the quantum discord of this system is not symmetric, we can prove that the minimum basis entropy of this state is nonzero, which means theorem 2 still holds. One thing we should note here is that, when the quantum discord is not symmetric, the minimum basis entropy is not exactly the same as the non-zero quantum discord. Take the above state (C.10) as an example, its quantum discord $\delta(A:B)_{\{\Pi_i^B\}} = 0.1887$, the measurement projectors are

$$B_0 = \frac{1}{2}\begin{pmatrix} 1 & 0 \\ 0 & 0 \end{pmatrix}, B_1 = \frac{1}{2}\begin{pmatrix} 0 & 0 \\ 0 & 1 \end{pmatrix} \tag{C.11}$$

while the minimum basis entropy of the joint state is

$$\begin{aligned} BE_{min} =& S[\frac{1}{4}\begin{pmatrix} 2 & 0 & 0 & 0 \\ 0 & 0 & 0 & 0 \\ 0 & 0 & 1 & 0 \\ 0 & 0 & 0 & 1 \end{pmatrix}] - S[\frac{1}{4}\begin{pmatrix} 2 & 0 & 0 & 0 \\ 0 & 0 & 0 & 0 \\ 0 & 0 & 1 & 1 \\ 0 & 0 & 1 & 1 \end{pmatrix}] \\ =& 1.5 - 1 \\ =& 0.5 \neq 0.2896 \end{aligned} \tag{C.12}$$

So, the minimum basis entropy of the joint system could be used as a detector of quantum discord, it would be inappropriate to use it to quantify quantum discord in some states.

References

[1] Nielsen, M. A., & Chuang, I. L. (2010). *Quantum computation and quantum information.* (Cambridge: Cambridge university press)
[2] Cover, T. M., & Thomas, J. A. (2012). *Elements of information theory.* (Hoboken, New Jersey: John Wiley & Sons)
[3] Bengtsson, I., & Zyczkowski, K. (2007). *Geometry of quantum states: an introduction to quantum entanglement.* (Cambridge: Cambridge university press)
[4] Grover, L. K. (1996, July). A fast quantum mechanical algorithm for database search. *In Proceedings of the twenty-eighth annual ACM symposium on Theory of computing* pp 212-219
[5] Lavor, C., Manssur, L. R. U., & Portugal, R. (2003). Grover's Algorithm: quantum database search. (arXiv *Preprint* quant-ph/0301079)
[6] Where $|\psi\rangle$ is the initial search state $(\sum_{i=0}^{2^n-1}|x_i\rangle)/\sqrt{2^n}$, **O** is the oracle, for it's definition, see [1], section 6.1.1
[7] For a joint system ρ^{AB}, the quantum discord of this system comes from $S(A|\{\Pi_B\}) - S(A|B)$, the conditional entropy $S(A|B)$ is increased by measureing B in orthogonal projectors $\{\Pi_B\}$
[8] Ollivier, H., & Zurek, W. H. (2001). Quantum discord: a measure of the quantumness of correlations. *Phys. Rev. Lett.* **88**(1), 017901
[9] Luo, S. (2008). Quantum discord for two-qubit systems. *Phys. Rev.* A. **77**(4), 042303
[10] Baumgratz, T., Cramer, M., & Plenio, M. B. (2014). Quantifying coherence. *Phys. Rev. Lett.* **113**(14), 140401
[11] Aberg, J. (2006). Quantifying superposition. (arXiv *Preprint* quant-ph/0612146)
[12] Ketterle, W., & Miesner, H. J. (1997). Coherence properties of Bose-Einstein condensates and atom lasers. *Phys. Rev.* A **56**(4), 3291
[13] Streltsov, A., Singh, U., Dhar, H. S., Bera, M. N., & Adesso, G. (2015). Measuring quantum coherence with entanglement. *Phys. Rev. Lett.* **115**(2), 020403
[14] Schlosshauer, M. (2005). Decoherence, the measurement problem, and interpretations of quantum mechanics. *Rev. Mod. Phys.* **76**(4), 1267
[15] Wehrl, A. (1978). General properties of entropy. *Rev. Mod. Phys.* **50**(2), 221
[16] John Von Neumann. (1955). *Mathematical foundations of quantum mechanics.* (Princeton, New Jersey: Princeton university press)
[17] Vedral, V., Plenio, M. B., Rippin, M. A., & Knight, P. L. (1997). Quantifying entanglement. *Phys. Rev. Lett.* **78**(12), 2275
[18] Van Enk, S. J., & Fuchs, C. A. (2001). Quantum state of an ideal propagating laser field. *Phys. Rev. Lett.* **88**(2), 027902
[19] Glauber, R. J. (1963). Coherent and incoherent states of the radiation field. *Phys. Rev.* **131**(6), 2766
[20] Glauber, R. J. (1963). The quantum theory of optical coherence. *Phys. Rev.* **130**(6), 2529
[21] Datta, A., Shaji, A., & Caves, C. M. (2008). Quantum discord and the power of one qubit. *Phys. Rev. Lett.* **100**(5), 050502
[22] Horodecki, M., Oppenheim, J., & Winter, A. (2005). Partial quantum information. *Nature.* **436**(7051), pp 673-676

Institute of Physics
Dirac House, Temple Back
Bristol BS1 6BE UK

ISSN: 1742-6588
ISBN 978-1-5108-3061-5

CURRAN ASSOCIATES INC.
proceedings.com

1st International Conference on Environmental Sciences (ICES2018)

Disaster Mitigation, Environmental and Sustainable Development

IOP Conference Series: Earth and Environmental Science
Volume 314

Padang, Indonesia
15 - 16 November 2018

IOP
Institute of Physics